THE MAN ON HACKPEN HILL

JS MONROE

HEAD of ZEUS

An Aries Book

First published in the UK in 2021 by Head of Zeus Ltd
This paperback edition first published in 2022 by Head of Zeus Ltd,
part of Bloomsbury Publishing Plc

9 7 5 3 1 2 4 6 8

A catalogue record for this book is available from
the British Library.

ISBN (PB): 9781838939595
ISBN (E): 9781789541700

Typeset by Divaddict Publishing Solutions Ltd.

Printed and bound in Great Britain by
CPI Group (UK) Ltd, Croydon CR0 4YY

Head of Zeus Ltd
First Floor East
5–8 Hardwick Street
London EC1R 4RG

WWW.HEADOFZEUS.COM

THE
MAN
ON
HACKPEN
HILL

In memory of Stewart McLennan

He who marches out of step hears
another drum

—Ken Kesey

Erin

I blame the rooks. For weeks they've been urging me to fly, hurling insults, calling me names. Corvid the coward! Featherless fucktard! Even worse, chicken thighs! Bella's right. I've been on it a lot lately. Anything to silence the birds in my head. Stop the heckling. The taunts. And now my best friend has left. The only person who understood, gave me time. Respect. Good luck to Bella. She's going to need it.

Haslam's still out front, talking to a couple of prospective parents, lying his skinny arse off. They have no idea what really goes on here. The pressures, the price young people pay.

A quick check up and down the corridor. Deserted. No surprise there. Picking open the buttons, I take off my shirt, followed by my jeans and underwear, folding them in a neat pile beneath the window. Another check and I'm up on the banister, steadying myself against the newel post. Bella and her mam had met here earlier, down on the polished wooden floor below.

My legs start to spasm and for a second I think I'm going to fall. But then I'm standing, both arms stretched above me in triumph. A naked fledgling about to fly the nest. I look up at my outspread wings. There must be an open window somewhere. My feathers are shimmering, ruffled by a warm summer breeze. And they are ready to carry me down to the hall below and out into the gardens, free at last.

I curl my toes over the banister, cock my head from side to side, and blink like a bird. I am a bird.

'Erin, come down from there.'

It's Haslam, standing by the front door. Calm, controlling. Not this time, pal.

'Erin, we can talk about this,' he says, his voice more urgent now.

I throw back my head and cry out like the rooks I will soon be among, high up in the sycamores where Haslam can't reach me.

'You don't have to do this,' the little bird behind my ear says.

Confused, I hesitate, keeping my balance on the banister. My whole life has been leading up to today. I've thought about it for months, planned it, dreamt how it will feel.

'I don't?' I ask, my voice barely a whisper. A tiny bud of relief unfurls inside me like a spring fern.

If the little bird answers, I can't hear it. The air all around me is torn apart by a hideous cacophony of rooks.

'Yes she does!' they cry as one. 'Yes she feckin' does!'

I look down at Haslam, racing up the staircase towards me, and then back at the rooks, who have darkened the sky as they swarm and squawk, mobbing the mullion windows.

'Erin!' Haslam shouts, nearly at the top now, raw panic in his voice. The sound of people behind me, running down the corridor.

I close my eyes, flap my wings – and fly.

1

Bella

Bella turns around to take one last look at the college as her mum drives away through the gates. A porter nods at them from the lodge. For the past three years, Bella's been studying at one of Oxford's smaller colleges, tucked away on the outskirts of the city. *An oasis of Victorian brickwork in a sea of manicured lawns and sycamores.* At least, that's how she described the place in an article for the college mag.

Her mum doesn't speak at first, which suits Bella fine. It's a bright June day and the outside world is overwhelming, as if the volume's been turned up too loud.

'You're looking so well,' her mum says, as they leave Oxford behind. She glances across at Bella, shaking her head in disbelief.

'You mean I'm not so fat,' Bella says, smiling.

In the past couple of months, she's shed all the weight she put on during her three years at uni. The pounds just seemed to fall away. She steals another look in the mirror. Erin trimmed her fringe last night, said she looked edgy with it shorter, but that was without her glasses. She takes them off, pushing back her long black hair, and leans into the mirror. She doesn't feel very edgy. In Freshers' Week, someone had called her a 'lanky librarian' and she's never forgotten it, even when she put on weight.

'Your skin also seems so… young.'

'Yours too, Mum.'

Bella peers over the top of her glasses. Her mum's skin is a delicious olive colour from her years of living and working abroad. Her eyes are tired, though; etched with grief. At least she's wearing a pretty bohemian dress, making an effort.

'And I like your hair,' Bella adds. It used to be ash brown but she's let it age naturally.

'Silver, not grey, OK?' her mum says, flicking it back.

'Got it,' Bella says, nodding. 'Can I put the radio on?'

'Sure,' her mum says.

It will be good to spend time together, help around the house and at work. During the week, her mum runs a local migrant centre near their home in Homerton, east London. At weekends she tries to stave off empty nest syndrome. Bella's elder sister, Helen, really did fly away – to live in Australia.

'Ready?' Bella asks, finger hovering over the radio.

Her mum glances across at Bella, her mouth creasing into a smile. Who can name the artist first? It's a game the family used to play when she was younger, driving down to Studland Bay. And Helen – happy, party-loving Helen – always knew the song, leaving Bella to sulk with a book.

'Jackie Wilson!' her mum shouts as the music comes on, joining in with the chorus. Bella sings along too, watching the Oxfordshire countryside slide by. No more sulks. It's a while before she realises her mum's crying.

'What's wrong, Mum?' she asks, turning down the radio and putting a hand on her shoulder. She knows the answer. It's been three years since Helen left for Australia and she's yet to return home for a visit.

'I'm sorry,' her mum says, sniffing. 'I'm just so happy to have you back.'

'You make it sound like I'm the one who's been living abroad.'

Bella notices her mum's hands tense on the steering wheel, knuckles whitening. The car suddenly feels hot and airless. Bella should have come home more, taken care of her mum. It must have hurt, having both daughters push her away. But it wasn't easy at Oxford. So much academic pressure, so little time to think of others.

'There's something you need to try to understand,' her mum says, setting her jaw. Her tone is unnatural, almost lecturing, which has never been her style as a mother.

'You fancy Dr Haslam,' Bella says, putting a plimsolled foot up on the dashboard. She's keen to keep things light, amicable. Dr Haslam was Bella's English tutor. Brain the size of the Bodleian and a PhD in Wordsworth and melancholia. Head of pastoral care too. And not for the first time there was an air of conspiracy between him and her mum when he came out to say goodbye today. Whispered asides.

'Please, he's nearly half my age,' her mum says, smudging away a tear with the palm of her hand. She glances at Bella's foot.

'That wasn't quite a "no", Mum,' Bella says, sitting up, rubbing a mark off the dashboard. She doesn't like this car – it's too small for her legs and *so* not electric – but her mum's driven a long way today. And Bella's promised herself to be less judgemental. Try not to point out how many lights her mum leaves on around the house, or dwell on the 2.5 tonnes of CO_2 generated by Helen's one-way flight to Sydney.

Her mum faces forward, concentrating on the road. Bella can picture Dr Haslam, running a hand through his long, swept-back hair as he takes her mum aside in his trademark skinny jeans and corduroy jacket, patched at the elbows.

'What is it, then?' Bella asks. 'We're driving straight to Studland for a weekend away?'

Bella regrets her words at once. She's done that a lot in recent months, said the first thing that's popped into her head, enjoying the speed with which her brain has started to make connections. It was after the last disastrous family holiday at Studland Bay that Helen decided to emigrate with her Australian boyfriend.

'You OK?' Bella asks, glancing at her mum again.

Something's very wrong. She knew things were too good between them to last. The next moment they veer off the road. Her mum jams on the brakes, bringing them to a shuddering halt in a dusty bus stop. They both sit in stunned silence, the car swaying like an unsteady drunk as a lorry thunders past.

'The f—?' Bella begins.

'Please. Just listen to me,' her mum interrupts, holding up a hand to prevent Bella from speaking. Her voice is loud and she's breathing hard. They both are.

'OK,' Bella says sarcastically, arms folded, staring out the window like she used to do as a child. 'I'm sorry. I'm listening.' It was tactless to mention Studland and maybe she went on a bit about Dr Haslam and she shouldn't put her feet up on the dashboard, but she can't help feeling her mum's overreacting.

'About your time at Oxford,' her mum continues, struggling to find the right words.

'It's over now,' Bella interrupts. 'And I apologise if I haven't been a very good daughter recently. I should have come back more often and—'

'Bella, it wasn't...' She hesitates.

'Wasn't what?' Bella prompts.

'It wasn't real life.'

Bella closes her eyes, breathing a sigh of relief that it's nothing more serious. Her mum's obsessed with the unreality of Oxford, always has been, ever since Bella was offered a place to read English. Its detachment from the rest of society, the isolation, the challenge of reintegrating into the normal world afterwards. The sheer elitism of the place. None of which is true of her own college, an establishment that prides itself on a high percentage of students from low-income families. It's anything but a bastion of privilege – just ask Erin.

Bella bites her lip, filled with a sudden, overwhelming sense that she might never see her best friend again.

'I'm well aware of that, Mum,' she manages to say. They'd hugged goodbye at the top of the stairs. At least, Bella had hugged Erin before she could recoil. Her friend's been in a bad way recently. Too many drugs and endless, all-night benders.

'You are?' her mum says, turning to Bella.

She nods, regaining her composure. 'Sure. We discussed

it a lot. With Dr Haslam. It's OK, I get it. Honestly. Real life starts now. That's what he said. And at twenty-one, I've got it all to play for. "*Bliss was it in that dawn to be alive. But to be young was very heaven!*" He used to quote that kind of cool shit – stuff – all the time. If it wasn't Wordsworth, it was Keats. And I'm ready to move on, I know where I'm going. What I want to do with my life.'

Unlike Erin. No one came to pick her up today.

'Can we drive on now?' Bella asks.

'Sure, I'm sorry.' Her mum checks the wing mirror, swallows hard and pulls out. 'Let's get you home.'

'Give me a month – OK, maybe three months – and you'll be reading my byline in the papers,' Bella continues.

She'll call Erin when she gets home. Check she's OK.

'You'll see it one day. I promise,' she adds.

'OK, sweetie,' her mum whispers, her voice so quiet that Bella can hardly hear her. 'Whatever you say.'

2

Silas

Two months later

DI Silas Hart doesn't feel comfortable being away from the office on a Thursday morning. He doesn't feel comfortable doing the cobra either – his lower back's killing him – but that's what the yoga teacher has asked him to focus on while the rest of the class, including his wife, Mel, do the bow pose, which is quite beyond him. As head of Swindon CID, he should be at his desk in Gablecross police station.

'You're doing great,' Mel whispers. She's lying face down on the mat next to him, still holding both ankles behind her back. How does she *do* that?

'I don't feel great,' he says. This is for her, one of the shared activities that the marriage counsellor suggested. Being his normal scratchy self would defeat the object of the exercise so he manages a smile. 'I'm glad we're here though,' he adds.

'Shhhh,' Mel says, still smiling.

'I feel more in touch – with myself, with us.'

'We'll have you touching your toes before you know it,' the instructor says, adjusting his posture.

'Didn't know I had any – been so long since I saw them,' he says, his hands beginning to shake on the mat under the weight of his body.

'You can lower yourself down gently now,' she says to him. 'Exhale with the movement.'

He blows out his cheeks like a steam train as his chest drops to the mat. Mel is still a picture of calm, poised in all her flexible glory.

'You look beautiful like that,' he whispers, breathing hard, face squashed against the mat. 'Sexy.'

'Shhhhh,' she says again. 'That's not what this is about.'

He means it, though. Mel's a new woman, retraining as a florist and relishing the challenge of a second career. They met twenty-five years ago when she was a nurse, a job she loved and loathed in equal measure, the long hours eventually outweighing the satisfaction of saving lives. The two of them are back living together full-time after a rocky few years.

Maybe it's a coincidence, but their twenty-three-year-old son, Conor, is a new man too, addressing his own psychological issues. The skunk-induced psychotic episodes, his disappearance for six weeks, the brief stint as a county lines drugs runner – they all seem distant memories now. Thank God. They are a happy, normal family again. If only Silas could shift the amount of weight that Mel's lost in recent months. He's not sure the yoga's going to do it.

Afterwards, as they make their way to the car, Silas's work phone vibrates in his pocket. He promised Mel he'd leave it in the glovebox but he's brought it with him, on silent. Should he answer it? The plan now is to drive over to Bath for a sauna and rooftop swim at the Thermae spa, followed by a light lunch of a few salad leaves. All part of their new life. Silas can't even begin to imagine how hungry he'll be feeling by dinner. The phone stops vibrating – and starts again.

'Damn, I've left my towel inside,' Silas says, turning to walk back to the leisure centre.

'Answer it,' Mel says, clicking open their car. 'You didn't bring a towel.'

She knows him too well.

'Won't be a sec.'

He raises his eyebrows in apology, as if it's not his fault. It must be DC Strover, his young colleague. She knows not to contact Silas on his 'date days', as Mel insists on calling them, but she also knows to ring him twice if it's urgent.

'Sorry to bother you, sir,' Strover says in her soft Bristol accent. He's told her countless times that he prefers 'guv' or 'boss' to 'sir', but she can't seem to make the switch.

'Luckily for you I'm "chilled", having just done an hour of hatha yoga,' he says, waving at Mel, who's now in the car, waiting for him. In the past, she would have bitten his head off for taking the call, but he's in credit after the yoga and she smiles back. She's a good woman, had to put up with too much from him over the years. 'What is it?' he asks.

'The boss has been over, wants you to be SIO on a new case that's just come in.'

Silas has a fluctuating relationship with their recently promoted boss, Detective Chief Superintendent Ward.

'And it can't wait until tomorrow?' he says, glancing at his watch. 10 a.m. His day off has hardly started. It must be a homicide for Ward to ask him to be senior investigating officer. 'What's the case?'

'A crop circle's appeared, below Hackpen Hill.'

Silas pulls a face at his phone, glancing over to Mel for sympathy, to let her know how ridiculous the call is, that it's not his fault he's been contacted on a date day about something so trivial. 'You are kidding me, Strover.'

It's Ward, getting his revenge by asking him to investigate aliens and angry farmers.

'A rambler found it,' Strover continues. 'It's a really strange pattern, no one knows what it is.'

'I don't care if it's the Shroud of bloody Turin, why's Ward giving it to us?'

'Because the rambler also found a dead body in the middle of the circle. A young man.'

Silas freezes. *A young man.* Could it be Conor? He's just being paranoid. Their son's better now. Admittedly, he was a bit edgy when he came home last night. Furtive. But that's because he said he'd be back for dinner and then came in late with Emma, his new girlfriend. It wasn't a problem. He'd texted them. So why was he so jumpy? Was he starting to feel unwell? It's happened before. They both know the signs.

He glances over at Mel, whose own demeanour has changed. She must have detected Silas's anxiety.

'The body was naked,' Strover continues. 'Looked like it had fallen from the sky.'

3

Bella

Bella stares at her screen, wondering which restaurant to ring on her list. Her first task this morning is to book somewhere for her boss to have lunch. It's not exactly going to win her the Pulitzer Prize. Maybe Dr Haslam was right and she's wasting her time in journalism. But she knows he's wrong. She's destined to be a journalist, wanted to be one for as long as she can remember. Just like her dad.

She glances over at her boss, who's on the phone. Mark's nice enough, in an old-school sort of way – likes his long liquid lunches, quick with the witty one-liners. He knew her dad from years back, when they were both foreign correspondents in Nairobi and her mum was working with an aid agency in Mombasa. It was her mum's idea to write to Mark and he replied at once, offering work experience.

A man from the post room stops by Bella's desk with a trolley and dumps a load of mail in front of her to distribute around the department. He winks and Bella manages a smile in return.

'Anything for my dinner?' he asks, sniffing the air. The food desk in the corner is always being sent free samples.

'Vegan custard?' she says, reading from the label of a tin on her desk.

'Rather stick pins in my eyes,' he says, starting to push his trolley away. She knows what he means. 'There might be a story in that lot, you never know,' he adds, nodding at the pile of post he's just delivered.

If only. In her original letter to Mark, she'd expressed her passion for the environment, how she hoped to write about green space and ghost nets, her desire to expose carbon crimes and climate criminals. Instead, she's sorting press releases on clementine sorbets and mood lighting. She looks again at the can of custard, takes a picture on her phone and sends it to Erin. Her best friend's a vegan and has tried to persuade Bella to follow suit. As a card-carrying veggie, Bella's all for veganism – if it wasn't for the invention of cheese. She just can't live without Brie. Or scrambled eggs, for that matter. And maybe the middle of a pain au raisin… Any patisserie, if she's honest.

Her tummy rumbles as she remembers a drugged-fuelled night in the college kitchens. Erin was making vegan custard slices for Bella in a final bid to convert her. Except that she'd overdone the turmeric and the slices were disgusting. Bella and Erin ended up throwing them at each other – just as a porter walked in. He almost took one in the face and didn't find it funny.

Bella smiles at the memory, turning the can in her hands. And then a tear rolls down her cheek. She hopes her friend's OK. Dr Haslam rang Bella on her first evening home from uni, to say that Erin had been taken very ill and was in hospital. He wouldn't say where or what was wrong with her, only that she was too unwell to receive visitors.

Two months later and Bella's still trying to get some answers. Despite her repeated requests, Dr Haslam won't tell her anything else – patient confidentiality, apparently – and he has started to ignore her messages. Bella has rung every hospital within a hundred-mile radius of Oxford and argued with her mum, who insists Erin's in good hands and just needs time to recover. She wipes away the tear, sadness replaced by anger. It makes Bella so vexed. What right has Dr Haslam to deny her access to her friend? He's aware how close they were at college, used to joke they were like sisters. Birds of a feather.

In his last email, almost a month ago, he'd promised to let Bella know when Erin was well enough to be visited. How kind of him. Erin's own phone keeps going straight to voicemail. Maybe she's too unwell to check her messages? It would be peak if she thought Bella wasn't concerned, had ghosted her as soon as she'd left uni. She's texted her so many times, tried to ring. If it were Bella in hospital, she'd expect her best friend at least to make contact.

She glances around the office as she sifts through the post, sorting piles of press releases. Bella's not ungrateful, knows how lucky she is to have work experience. On a national newspaper too. For the first two months at home, she helped her mum at the migrant centre, went for long walks in Victoria Park, adjusted to life outside college, talked about Erin. Healing time, her mum called it.

'Hold up, darlin', this one's for you,' the post man says, glancing at Bella's name on the envelope as he returns with his trolley.

'For me?' Bella asks, flattered that someone knows her name. She takes the letter and opens it. It's been written on an old typewriter – 'always a bad sign', Mark had said when he was briefing her about readers' letters, the vast majority of which come via email, but Bella's impressed that someone's gone to such trouble. She's always preferred sending old-fashioned letters, particularly to Helen.

The newspaper has a column called 'Overheard' about community pubs and the 'loyal reader' is suggesting that it should feature a place called the Slaughtered Lamb in a village in Wiltshire. The column comprises snippets of bar chat and the reader insists there's lots of 'seriously interesting stuff' to be overheard at this particular pub.

Bella puts down the letter, thinking about the pub's gruesome name. Why not call it the Pampered Lamb? Meat is the one thing she knows she and her dad would argue about if he were still around. Apparently, he loved nothing better than tucking into a steaming bowl of beef suqaar, his favourite Somali stir-fry.

Bella assumes the letter has been sent in anonymously by the landlord as a way to get publicity. But why's it addressed to her? She's yet to have a byline in the paper. She rereads the letter, noticing a PS on the back: 'Make sure you "overhear" the man sitting on his own in the corner.'

Will he be wearing a red carnation too? It's not exactly subtle. She checks the envelope again. No postmark. Perhaps it's from her mum, matching her up with someone. She's always going on about Bella's need to find Mr Right. Or maybe she's trying to get Bella out of the house so she can have time on her own with Dr Haslam.

She glances around her colleagues and googles the pub.

The Slaughtered Lamb is in a village an hour on the train from London. She could go down there at the weekend. Nothing to lose and maybe the only way she'll get a byline around here. Eavesdropping is right up her street too. She's also been looking through her dad's old journalism cuttings. In the early years, before he started filing front-page exclusives from Africa for *The Washington Post*, he wrote about pension relief for an accountancy magazine in Bracknell. Everyone has to start somewhere.

She gets up from her desk and checks her watch. 11 a.m. in the UK, 9 p.m. in Sydney. Helen will be out partying, won't wait to chat. Standing at the window, she looks down on the street below. A normal summer morning in central London, tourists mingling with office workers on the coffee run. And then her eye is caught by someone, talking to another man in a shop doorway. He glances up in her direction. His face is in shadow but she'd recognise that corduroy jacket anywhere. Dr Haslam.

4

Silas

Silas pulls over at the top of Hackpen Hill and gazes down on the golden wheat fields below. Crime Scene Investigators in white body suits are moving around the centre of the crop circle, where a tent has been erected. As soon as he was notified about the body, he put in a call to the coroner, asking for a Home Office forensic pathologist to attend the scene. His old friend Malcolm is on duty and Silas has already texted him a photo of Conor. Malcolm replied a few minutes ago, shortly after arriving on site. He doesn't think the dead man is Conor, thank God.

'Jesus,' Silas says, as he and Strover get out of the car and look down on the field below them. The crop circle comprises two distinct hexagons and various other geometric shapes, and a much larger – and dramatic – spiral-patterned disc beside it. In total, the pattern must be about 100 metres long and 50 metres wide. 'Someone's been on the cider,' Silas adds.

Crop circles are a pain in the arse, as far as he's concerned. Dreamt up in the pub over a pint or three by people with nothing better to do and then hailed as extraterrestrial messages by the croppie community – the believers.

'Isn't it incredible?' Strover asks.

Silas glances at Strover. He'd picked up his junior colleague from Gablecross after yoga and headed straight here. Mel was understanding. They'll go to Bath another day.

'"Incredible" isn't the first word that springs to mind,' Silas says. 'What you're looking at is a criminal offence that causes damage to crops and a loss of revenue to the farmer.' His words hang in the summer air. If Mel were here, she'd be telling him to be more open-minded and less grumpy. 'You know each circle costs the farmer up to a grand in lost revenue,' he continues without much conviction, looking again at the intricate patterns spread out below them. 'The combine can't pick up the flattened wheat.'

Strover remains silent. She doesn't need to speak. They both know that this circle is different from the ones that have been appearing recently. It's spectacular.

'What's it about, then?' he asks grudgingly, as his eyes trace the beautiful swirling patterns where the wheat has been compressed. 'ET trying to phone home?'

Strover shakes her head, deep in thought.

'I thought you were good at puzzles,' Silas adds.

'There's quite a history of this type of circle around here,' Strover says, eyes still fixed on the field below. 'Coded messages, complex mathematical equations.'

It's the last thing he needs. The media will be all over this in no time. As senior investigating officer, he should be down at the crime scene now, but he wanted to come up here first to get a feel for the setting, try to understand the unusual context. He's aware that some crop circles can only be 'appreciated' from the air, but this location is visible from the road that winds down the side of Hackpen Hill, a scenic back route from Marlborough to Royal Wootton Bassett. Whoever left the body in the middle wanted it to be spotted by drivers. It's also overlooked by a popular local landmark – a sleek white horse carved into the hillside.

'Someone from the Crop Circle Exhibition and Information Centre will be here shortly,' Strover adds.

'Sign of the times,' Silas says. A truckers' café on the road to Calne used to double up as the croppies' HQ during summer. Much to Silas's dismay, Wiltshire has become the epicentre for crop circles in the past thirty years – as many as a hundred can appear in a summer – and it's now a global business. Two years back, a Chinese movie star arrived in Wiltshire with a forty-strong entourage to open the Exhibition and Information Centre at Honeystreet, later chartering a helicopter to fly over a nearby crop circle for a prime-time TV show back home.

'They're very concerned,' Strover says. 'There's never been a death in one.'

'Just tell me it's got nothing to do with novichok,' Silas says. 'Social media's awash with possible Russian involvement.'

'Didn't know you were on social media, sir?' she asks, suppressing a smile.

It's true, Silas has left social media to her in the past, but he's recently taken to it to follow the horses.

'I've been known to twitter,' he says.

The last thing he wants is the Met's SO15 swanning down from London again. Four months after the original novichok attack in Salisbury in 2018, a crop circle had appeared outside Amesbury in Wiltshire, a day before a local woman called Dawn Sturgess had died from the nerve agent. She'd inadvertently sprayed novichok on herself, after her partner had found what he thought was a discarded perfume bottle in a charity shop bin. The crop circle depicted the international symbol for chemical weapons, prompting conspiracy theorists to have a field day, as it were.

Wiltshire police should have investigated the case, but SO15 had taken over from the 'carrot crunchers', just as they'd done in Salisbury earlier.

'The specific novichok compound used in Salisbury is known as A234,' Strover says, reading from her phone, 'and has the chemical formula $C_8H_{18}FN_2O_2P$. Whether that equates to that,' she says, nodding down at the pattern, 'is anyone's guess.'

'Christ, let's hope not.' Silas knows Strover is good at maths. She's clearly useful at chemistry too.

'We can't rule out Covid-19 either,' Strover adds. 'All those protein spikes would make a great pattern, but I can't see it in that thing.'

'You're in the wrong business,' Silas says.

The world is still recovering from the pandemic, the personal tragedies, the recalibration of everyday life. Mel lost her mother, who'd always been a fan of theirs as a couple, kept them talking during the bad days, persuaded Silas to go to counselling. They're both making an effort for her.

'I hope that's one of ours,' Silas says, gesturing at a drone hovering over the site. It's almost at the same level as them on top of the hill.

'CSI are taking aerial photos,' Strover says, as the drone's blades flash in the morning sunshine. 'The crop circle appeared during the night. We're assuming the body did too.'

5

Bella

Bella tries not to run through the open-plan office as she heads for the escalator at the entrance. Cars hoot at her as she zigzags through the slow London traffic to the opposite pavement, where she stands, hands on knees, breathless. Dr Haslam has vanished, but the man he was talking to is still there, on his phone, turned away from her.

'Excuse me,' Bella says, stepping around to face the man. 'Sorry to bother you, but was that Dr Haslam you were talking to just now? From Oxford University?' She needs to ask him about Erin.

'Sorry?' the man says, putting one hand over his phone. Has Bella seen him before? Something about him – his hunched shoulders, jet-black hair, sharp suit – is familiar.

'Was that Dr Haslam? Just now? Right here?' Bella asks again. 'Talking to you?'

'I'm sorry, do I know you?' he says, his tone more aggressive now.

'Were you with Dr Haslam?' Bella repeats, her own voice growing louder. 'Fuck sake, just tell me. Please?'

Passers-by stop to look. Others give her a wide berth. She shouldn't have sworn, lost control. The man shakes his head, eyes still fixed on Bella's. Does he know her?

'I've never heard of him in my life,' he says.

The man walks away and glances back, almost as if he's afraid of her. She stands on the pavement, watching him disappear into the crowds, and looks around again, into the nearby shops, down the street in both directions. She should have been a better friend to Erin, not left her on her own at college. The signs of drug abuse had been there for a while. Her apathy, short attention span, wanting to withdraw from the world. But Bella had been too fixated on herself, losing weight, passing her final examinations.

'I wish you weren't leaving, like,' Erin had said on that last day, her Dublin brogue more pronounced. It always was when she was upset. She was slouched on a beanbag in the corner of Bella's college room, watching her take down sheets of paper from the walls – funny headlines, overheard quotes around college, story ideas.

Bella looked over to her, surprised. Neither of them had been very good at expressing emotion in their time there. Small notes slipped under each other's doors asking if they're OK, but seldom open gestures of affection. She knew she would miss Erin too, her tartan pyjamas and tattoos and crazy stories of busking with her dad in Dublin. It made Bella's own upbringing feel so sheltered.

'I'm only going to London,' Bella said, picking up her overstuffed rucksack and a bulging blue Ikea bag. 'Mum wants me to stay at hers for a few months – until I find a place of my own.'

'Sounds nice.'

There was no envy in Erin's voice, no self-pity, but Bella felt bad.

'Your mam's here already, by the way,' Erin added. 'Talking to Haslam.'

'You sure?'

Erin nodded, cocking her head sideways. Bella was about to ask her why she hadn't told her sooner, but she knew. She didn't want Bella to leave.

'Haslam's laying on the charm with a shovel,' Erin added. 'Better get down there quick before you have a wee sister.'

Bella looks up and down the London street, smiling at the memory. Despite their different backgrounds, they seemed to click from day one. Erin taught Bella to stand up for herself, to unbutton a little. And Bella went along with it, particularly the drugs in her first couple of years. It's what students did at uni, after all. Anything to shake off the lanky librarian moniker. But in the end, she wasn't brave enough to be like her friend. Erin's defiance of authority, the college porters, Dr Haslam. Bella chose to acquiesce. To study hard. And now she's left her behind.

6

Silas

Silas and Strover get back into his car and set off down towards the scene of crime, past the white horse. At the bottom of Hackpen Hill, they are waved through a police roadblock and park up beside the field. In the distance, on the road to Royal Wootton Bassett, Silas can see a convoy of camper vans congregating behind another barrier.

'Make sure those croppies don't get any closer,' he says to a uniform, gesturing down the road towards the vehicles. He's had dealings with croppies in the past, usually after complaints from farmers, and isn't in the mood for a long conversation about ley lines and why his mobile phone might suddenly stop working in the middle of a crop circle. He hasn't got the energy – unlike the circles themselves. As any self-respecting croppie knows, they are notorious for interfering with electronic equipment.

Silas and Strover don white disposable suits and shoes, sign the crime scene attendance log and enter the tent. He's seen plenty of dead bodies in his career but it still shocks him every time he encounters one. The day it doesn't will be the day he retires. The young man lying on the flattened wheat is in the centre of one of the hexagons, and for a second Silas wonders if he really has fallen from the sky. Everything about the scene feels staged. The body is on its side, almost as if in the recovery position, and seems peaceful. It's not Conor but Silas's relief is tempered by the knowledge that another father will soon be grieving.

Silas peers closer. It's the victim's face that most interests him. His complexion is pallid, as Silas would expect, but he appears to have two black eyes. Something about them is not quite right, though: they are darker at the top of the socket than below, where there is no bruising. And then he notices the left wrist. It's been lacerated but the cuts don't appear fresh. Nor is there any evidence of blood on the flattened wheat below him.

'Don't even begin to ask me the cause of death,' a familiar voice says behind Silas. It's Malcolm, his old friend and forensic pathologist, who has followed them into the tent. Malcolm has been in the business for thirty years, his manner an odd mix of posh academic and car mechanic. He should have retired by now but he lost everything in a messy divorce and needs the money. Possesses an impish sense of humour too, but not now, not here.

'Before you ask, I'd say he died elsewhere from a severed radial artery and was subsequently brought here,' Malcolm begins, matter-of-factly.

'How long after he died?' Silas asks. Pathologists hate being asked time of death. Unless the victim's wearing a watch that stopped at the exact moment they died, they'd rather not say, which is infuriating, given it's the first thing that an SIO wants to know.

'Impossible to say. The body temperature is unnaturally low. I've got a funny feeling about this one.'

Silas has got a funny feeling too. And he hates funny feelings. 'What about the two black eyes?' he asks.

'You noticed them?'

Silas nods.

'You'll be putting me out of a job,' Malcolm continues, smiling. His own eyes are deep set, his grey eyebrows bristling and unwieldy. 'I hope they're paying you well.'

'They look unusual,' Silas says, ignoring the comment. If they start talking salaries, they'll never stop. Malcolm's always going on about his workload, how there are not enough forensic pathologists in the UK any more. 'Do we know what might have caused the bruising?'

'Too early to say, but they're definitely irregular.'

'How do you mean?' Silas asks.

Malcolm turns to grab Silas by the shoulder. Age has stooped him but he's still a tall, imposing man.

'If I were to punch you in the eye,' he says, holding his clenched fist inches from Silas's face, 'you would expect burst capillaries and bruising in the fatty tissue all around the socket. But in this case, the only haemorrhaging is above the eye, which, as you rightly say, is unusual.'

'So what could have caused them?' Silas asks, pleased that Malcolm has let him go. He can get carried away, theatrical. Loves his amateur dramatics. And cricket. Never ask him about cricket.

Malcolm looks around, checks that no one else is within earshot. Strover is still with them.

'She's with me,' Silas says, sensing Malcolm's reluctance to talk in front of her. He's not sure whether it's because

Strover's young or a woman. Probably both. Malcolm can be a bit unreconstructed at times, makes Silas seem positively woke.

'In the 1940s and 50s, psychosurgery was all the rage,' Malcolm says. 'Operating on the brain to cure mental illnesses. It was pretty crude, barbaric stuff. Suspected schizophrenia? Sever the prefrontal cortex.'

'A lobotomy, you mean,' Strover says.

Malcolm nods, wrong-footed by Strover's intervention. Silas is pleased. He's trying to encourage her to be more confident, speak her mind. 'The frontal lobes – personality expression, will to live, organisation of thoughts, moderating behaviour, that sort of thing – were effectively destroyed,' Malcolm continues. 'At first they went in through holes bored into the side of the head. But then they discovered they could access the brain less intrusively by going in through the back of the eye sockets.'

Silas closes his own eyes. He dreads where this might be heading.

'I'll spare you the details,' Malcolm continues. 'Let's just say it involved a sharp surgical instrument called an orbitoclast and a mallet.'

'Thanks for that,' Silas says. For the first time in weeks he doesn't feel hungry.

'Known as a transorbital lobotomy,' Strover chips in.

Both men turn to her in surprise. Silas doesn't even want to guess how she knows about such things.

'Bedtime reading,' Strover adds, flashing a thin smile at them both.

'Developed by a physician in the States called Walter Freeman,' Malcolm says, regaining the upper hand. 'He used to practise his technique with an ice pick and a grapefruit.'

'And you think our victim might have been given one of these…' Silas asks, keen to move on.

'A transorbital lobotomy?' Malcolm says, still looking at Strover with curiosity. 'Quite possibly. It's a while since I've seen one. I'm pleased to say psychiatry has progressed. We just need to establish if it was carried out before or after the poor man slit his wrist.'

7

Bella

'I think I saw Dr Haslam today,' Bella says, sticking another page of her notebook up on the wall of her bedroom.

'In town?' her mum asks, leaning against the doorway with her arms folded. It's early evening and sunlight is streaming into Bella's book-lined room, which has remained unchanged since she went up to Oxford. A map of Africa above her desk in the corner, snapshots of old school friends on the wall beside her wooden bed from Mombasa. A framed photo of her dad on the dressing table.

'At least, he looked just like Dr Haslam.'

Bella has begun to doubt it was her tutor, but the man she confronted on the street... She'd definitely seen him before.

'It can't have been him,' her mum says.

'Why not?' Bella asks, turning around. Her mum's wearing one of the silver African necklaces that Bella's dad gave her, strung with big amber beads.

'Because he rang me this afternoon from Italy, wondering how you were,' her mum says, fingering a bead.

'Rang you?' Bella asks. 'I've been trying to speak to him for bloody weeks.'

'I know. He's concerned – worried you've taken the news about Erin very badly.'

'How else was I going to take it?' Bella snaps. 'She's my best friend.'

'Erin's going to be in hospital for a long time, Bel.' Her mum's tone has changed, become more serious. 'Dr Haslam will contact you as soon as you are allowed to visit her. He's promised me. In the meantime, it's best you stop ringing him. And all those hospitals.'

Bella turns back to sticking sheets of paper up on the wall. She wonders if Erin has received any of her messages. And then she thinks about Dr Haslam again. Short and gnomic, but with an unmistakable air of authority. Strange that he rang on the same day she thought she'd seen him. Is he really in Italy? He was always talking about visiting John Keats's grave in Rome, located outside the city walls because the poet wasn't a Catholic.

'I just don't understand why I can't visit her,' Bella says, tears welling as she rips out another page of her notebook. 'It's been two months.'

'No one can, sweetie. She's very poorly.'

Her mum walks over to hug Bella as she starts to sob. They stay like that for a while, her mum stroking Bella's hair, the smell of jasmine perfume faint but familiar. The smell of her childhood. Of happier times. Bella looks across at her desk, where there's another framed photo of her dad in his beloved Somalia. Helen on his shoulders, her in his arms. His hair is long and flowing, and he's wearing studious glasses.

Her mum follows Bella's gaze and her hug tightens. Her dad died when Bella was eight and Helen was ten. She never had the chance to know him adult to adult, like she has with her mum.

'You lit up his life,' her mum says. 'He was so earnest until you two came along.'

Her mum's being kind. Helen was the sunny one.

'What are all these anyway?' her mum asks, breaking off to walk around the room, reading the pages on the walls.

'Ideas for headlines,' Bella says. 'It's just the way I work. I like to see everything laid out.'

'"The Joy of Ceps"?' she says, pulling a funny expression.

'Foraging for mushrooms.'

'He would have liked that one.' She pauses before continuing. 'And he would have been so proud of the way you're following your dreams. When you and Helen were very young, in our house in Mombasa, you used to sit in the attic at two separate tiny desks, the old wooden ones with inkwells. Your dad had made a sign above Helen's that said "Big Cheese Editor", and one above yours that said "Intrepid Reporter".'

Bella's memories are faint but she can picture Helen at her desk. Or is she recalling a photo?

'I've been thinking a lot about Dad since I came down from Oxford,' she says. 'I want to do something for him. Make him really proud by breaking a big story. One that counts. It might be environmental, but it doesn't have to be.'

'I'm sure you will,' her mum says, but she doesn't sound very convinced.

'I've been going through more of his newspaper cuttings,'

Bella continues. 'Human rights abuses in Zimbabwe, government corruption in Somalia. Big pharma trials in Kenya. Illegal oil drilling in Nigeria. And there I am filing press releases for bidets.'

'You're just starting out,' her mum says. 'And it wasn't always easy.'

'For you?' Bella asks. She loves to hear stories of her childhood, even if they're not all happy ones.

'For us. As a family.' She pauses, fingering the amber necklace again. 'Sometimes I feared he'd never come back.'

Bella remembers the long periods her dad was away on an undercover story, the empty place at the kitchen table. The night before he left, he would always tell Bella and Helen that he'd be gone by the morning. Bella would fall quiet but Helen would quiz him, ask if he was going to wear secret disguises.

He was more playful than usual the last time he went away. 'Maybe I need a false moustache,' he said in a Poirot accent, grabbing one of Helen's crayons and drawing a ridiculous line across his upper lip. 'How do I look?' He puckered his mouth as Helen rolled around laughing on the bed. 'So, you know who it is, do you? Hmm, maybe I need a beard too,' he added, scribbling all over his chin.

When the laughing stopped, he looked across at Bella on her bed.

'Are you going to come back, Daddy?' Bella asked quietly, twisting her tiny fingers in her lap.

'Of course,' he said, coming over to sit next to her.

'Promise?' Bella asked, looking up at him.

'Promise. Haven't I always come back?'

It was true. He did return, usually with presents, not

expensive ones but toy prams and hula hoops made by local kids out of wood and wire, and bought on the roadside.

A hint of a smile broke across Bella's face as she contemplated her dad's silly beard.

'Can I have a disguise?' she asked.

'Of course you can,' he said, cradling Bella's serious face in his hands as he kissed her forehead.

A minute later, her cheeks were a riot of squiggles and twirls.

'There we go,' he said, admiring his handiwork. 'Mummy won't recognise you at breakfast.'

'You OK?' her mum asks, breaking into Bella's thoughts.

'Fine,' she says, but it's obvious that she's not.

'Come here,' her mum says, arms outstretched, as loud African funk music starts to boom out from downstairs. It's the Dur-Dur Band from 1980s Somalia, recently reformed. Her mum rolls her eyes. She's renting the front room of the house to lodgers, a pair of handsome Somali refugees who are helping out at the migrant centre that she runs.

'We've got two choices,' she says, holding Bella's hands. 'Either we go downstairs and tell them to turn it down...'

'Or...?' Bella asks, knowing what the answer will be.

'We join them and dance!'

Her mum loved to dance when they were little, usually with Helen. Bella had needed more persuading. Not any more. She follows her mum downstairs but as she passes the landing window, something catches her eye in the street below. She stops to look. A car has pulled up and a man gets out. She only sees him for a second before he disappears but she's sure it's the same person she confronted on the street today. Same jet-black hair and hunched shoulders.

'What are you waiting for?' her mum calls up the stairs. Bella lingers a moment longer at the window. Maybe she's wrong, imagining things.

'Coming,' Bella says, but her feet are no longer quite so keen to dance.

8

Silas

'No one can say we haven't got the best brains working on the crop circle pattern,' Strover says, coming off the phone. 'That was a professor at the Department of Pure Mathematics and Mathematical Statistics at Cambridge.'

'Does he have any idea what it means?' Silas asks, sitting back in his chair. They are in their favourite corner of the Parade Room at Gablecross, where the talk is of nothing else. The crop circle killing is already headline news and TV crews are parked up outside the police station, waiting for an official statement.

'It's a "she",' Strover says, throwing Silas one of her looks. 'And she doesn't know yet but she seemed to relish the challenge when I briefed her. Apparently, it's in a different league to previous coded patterns. A unique combination of mathematics and chemistry.'

'Meaning it will take time,' Silas says, sighing. It's the last thing he wants to hear.

'I'm setting up a Zoom call with her and an associate

professor of chemistry at the Molecular Sciences Research
Hub at Imperial,' Strover adds.

Silas hates Zoom, the difficulty of reading someone's
expressions and body language.

'Can't wait,' he says.

It's been busy since they returned from the crop circle
at Hackpen Hill and he knows that they need all the help
they can get. And not just because of the media interest.
The boss is on their case too. The golden hour had long
passed by the time the body was found, but Silas has been
on the front foot all day, aware that time is still of the
essence. He's been pushing the CSI manager for any leads,
however small, while trying to fast-track a DNA sample.
He's also called Malcolm every half hour for news on the
autopsy, while Strover's chased boffins. She's on a mission
too, fired up by the case's unusual semiotics. Her word,
not his.

She picks up the photo that's just spewed out of the
printer and passes it over to Silas.

'That was done in 2010, in a field near Wilton Windmill
in Wiltshire,' she says.

'What is it?' Silas asks, looking at the complex pattern,
a circle divided up by twelve radial spokes, each one
sprouting short concentric spurs, and carved out of vivid
yellow oilseed rape. The overall effect is not dissimilar to
twelve sails of a giant windmill.

'It's a binary representation of Euler's Identity – one of
the most beautiful and profound mathematical equations
known to man. Apparently.'

Silas looks up at Strover and then at the photo. 'Binary?
That?'

Strover seems genuinely puzzled by Silas's response. In keeping with the police's new recruitment policy, she is a graduate, studied forensic science at the University of the West of England in Bristol.

'You can see here that each spoke has up to eight little arcs coming off it,' Strover says, pointing at the photo. 'And eight's quite a significant number when it comes to binary and computers. Does eight-bit extended ASCII code mean anything to you?'

Silas shakes his head.

'You know how computers work in binary – one is on, zero is off,' Strover says, not waiting for an answer. 'Well, ASCII is a character encoding standard that was developed in the 1960s and converts those binary numbers, or bits, into more familiar text – letters, symbols and numbers.'

She points again at the photo of the crop circle pattern. 'When there's a short arc coming off one of these main radial spokes, that represents a one. Where there's a gap, an arc missing, that's a zero.'

'I'll take your word for it,' Silas says. He wishes he'd paid more attention at school.

'While we're at it, a string of eight bits make a byte,' Strover says. 'And some people reckon this circle represents how data is arranged on a computer hard drive.'

'Can we move on?' Silas asks. He should be home by now, having dinner with Mel, like any ordinary couple.

'Sure,' Strover says, but she's hit her stride. 'This particular crop circle attracted a lot of interest from mathematicians around the world. One pointed out that when the binary was translated into text using ASCII, what should have been an "i" had become "hi", which was either someone's idea of a joke or a reference to Planck's Constant, the elementary quantum of action that's depicted with an "h".'

'Planck being...?' Silas asks, thoroughly baffled.

'Max Planck. A German theoretical physicist. And a reference to planks of wood, of course, which we all know are used by the people who make these crop circles,' Strover adds.

'Of course. Very funny.'

In a niche sort of way. It's not exactly Silas's sense of humour. He prefers his jokes more earthy.

'This one caused an even bigger stir,' Strover says, taking another photo off the printer. 'Possibly the most complex crop circle ever seen in Britain. Until now.'

9

Bella

Bella lies in the darkness, listening to the sounds of east London. It feels strange to be back in her old room, as if her time at Oxford never happened. The luminescent stars that her dad once gave her in Mombasa are still stuck on the ceiling, glowing faintly, a fading reminder. After his death, they moved back to the UK and this house in Homerton. Her mum had bought it cheaply in the 1990s when she was left some money. In those days, crime rates in the area were high – Clapton Road, round the corner, was known as 'Murder Mile' – but it's become gentrified in recent years. And the schools are getting better. Bella went to Clapton Girls' Academy, where she was one of several students in her year to win a place at Oxford, but she's drifted away from all her old friends.

A shame, given she'd like to take someone with her when she visits the Slaughtered Lamb. Erin would shake things up in a genteel pub in rural Wiltshire – and blow Bella's cover as a journalist. She'd be useless! Her friend wasn't really into pubs or drinking at uni. Being out of her head on drugs in her room was more her thing. If she came along with Bella, she'd be asking the barman for Q-balls.

A siren wails up the Chatsworth Road. Did an ambulance come for Erin? Bella can't stop thinking of what might have happened to her after term finished. Another lonely bender but this time more serious, trying to block out the pain of her past. She was never hospitalised before. Erin finally slipped through the cracks. Haslam went on about the importance of pastoral care but the number of students who dropped out because of drugs and mental health issues was shocking.

On one particularly heavy night in her first year, Bella was in Erin's room and both of them were pretty dosed. Erin was on her bed, knees drawn up, head rocking. Bella lay flat on the floor, room spinning like a dervish.

'Did you really sleep rough on the streets in Dublin?' Bella asked, trying to get her bombed brain around Erin's childhood.

'With my da. Sure we did. In the summer. Danced all day, happy to sleep anywhere, I was so knackered.'

Erin used to busk for the tourists on Grafton Street, step dancing to her dad's Irish fiddle playing, but she didn't like to talk about it. She was self-conscious, no longer the cute slip of a girl that used to pull in the crowds.

'How old were you?' Bella asked. 'When he died?'

'Ten.'

The same age as Helen when their own own dad died. Maybe that's why they got on so well. They had little else in common but sometimes it only takes one thing for a friendship to form.

'We were sleeping in a tent in Phoenix Park,' she said.

'I tried to wake him – the Gardaí were outside, arresting people. It was never easy in the mornings, because of his boozing, but this time…'

'Did you see him?' Bella asked, propping herself up on one elbow. 'His dead body?'

'What kind of a question's that? Sure I saw him dead. We were sharing a feckin' tent.'

Bella closed her heavy eyes and lay back down again, feeling dizzy.

'I'm sorry,' she said. 'But I envy you. The certainty of what you saw.'

'You're more out of your box than I thought,' Erin replied.

'It took me years to accept my dad was dead,' Bella said.

More silence, except for a lone rook outside, disturbed by something in the night. 'Did you get on with him OK, like?' Erin asked.

An image of Bella's dad, dancing around the kitchen table with her mum in Mombasa. 'I loved him,' she says.

Erin didn't speak for a while. And then Bella heard her sobbing. She'd never known Erin cry before. Or show any real emotion. Bella stood up, unsteady on her feet, and trudged over to her bed.

'It's OK,' Bella said, slinging an arm around Erin as she flopped down next to her, the room still spinning.

'I loved my da too,' Erin said. 'Overlooked all the shite stuff. We do, don't we? As daughters.' Bella nodded. She was sure that her own dad had lots of faults but she couldn't think of any right then. 'He kicked my mam out when I was six – after she'd tried to kill us both. I never saw her again. She was drunk and gave him a proper beating – he wouldn't hit her back. And then she turned on me with a kitchen knife and he knocked her clean out. I miss the stupid fella every day.'

10

Silas

'It's a geometric representation of the first ten digits of pi,' Strover says.

Silas shudders at the mention of pi. He rotates the photo in his hands, hoping to make better sense of the pattern of concentric circles. It's like a cross between an archery target and a spiralling maze, with three smaller circles floating on the perimeter, flat and smooth, echoing the bullseye.

'Found in a field of barley in 2008, near Barbury Castle,' Strover adds.

'Not so far from Hackpen Hill,' Silas says, thinking back to the naked body in the field, the bruised eye sockets. The Parade Room is almost empty. Most sensible people have gone home to their families by now.

'No one knew what it meant until an American called Mike Reed, an associate professor of astrophysics, made the link,' Strover continues. 'Each angular segment represents a digit. That little dot here' – she points to a tiny circle next to the centre – 'represents the decimal point. The tenth digit's even been rounded up properly.'

'Interesting,' Silas says, bluffing. God, how he hated maths at school. 'Not made by time travellers then.'

'Sorry?' Strover asks.

'Didn't you know? They leave crop circles as navigational aids,' Silas says. 'Just one of the many batshit explanations I've been given today for how they're formed. Along with plasma vortices, balls of light, ley lines and UFO landings.'

If the case wasn't so urgent, the theories would be funny. But Silas needs to find out who created the circle at Hackpen Hill and establish if there's any link with whoever placed the body at its centre. Unfortunately, the crop circle world is not so simple. The people – and Silas is in no doubt that it *is* people – who make these things don't like to break cover, not for fear of upsetting the croppies – or cereologists, as the believers also call themselves – but because they might get sued by the farmers for trespass and criminal damage. Having said that, some of the more savvy farmers cooperate with the croppies, charging the public for access to circles that appear in their fields.

Silas's phone rings. It's Malcolm.

'The body's been in a morgue,' Malcolm says. 'Before it was left in the field.' Silas likes it when people cut to the chase, but he's not so sure in this case. His life has just got a whole lot more complicated.

'Are you sure?' he asks.

'Some of the internal organs were still frozen,' Malcolm continues. 'It's been kept in a negative temperature storage facility – which usually run at minus 50 to prevent decomposition. If you want a speculative timeline, and I know you always do, I'd say he died from a slash to the left wrist, most probably self-inflicted, at least a month ago. Maybe longer. Within the past twelve hours, the fibrous matter that connects the cortical tissue of his brain's prefrontal cortex to the thalamus has been transected. The procedure – a transorbital lobotomy, as your young colleague rightly said – was carried out by accessing the victim's brain through the top of both eye sockets.'

'Hence the black eyes.'

'Correct.'

'Why would someone do that?' Silas asks, thinking aloud. 'Perform a brutal operation that was discontinued more than seventy years ago?'

Strover looks up.

'That's for you to find out, not me,' Malcolm says. 'A transorbital lobotomy is quite a specialist procedure, not immediately obvious to the layman. Although you did spot the bruised eye sockets.'

Silas signs off and pushes back on his chair, checking his mobile. A text from Mel, asking if he's coming back for dinner. No chance. He must remember to reply.

'Oh Christ,' he sighs, before filling Strover in on what Malcolm's told him.

Silas's phone rings again. The National DNA Database. He takes the call, listens, hangs up.

'No matches.' Strover's already checked the victim's face against the UK custody image database and his fingerprints have been run through IDENT1 and IABS, the criminal and immigration databases. The Vulnerable Persons Database and Missing Persons DNA Database have also come back with nothing.

'Our man who fell to earth is a ghost,' Silas continues. 'We need to put a message out to every mortuary in the country, see if they've had any slit-wrist suicides in the past two months and then ask them to check their chillers. Make sure no cadavers have gone walkabout.'

11

Bella

The phone rings downstairs. Bella's mum takes the call and walks back up to her bedroom, talking quietly. Maybe she has got a new man after all. Bella hopes her mum finds love again one day. She gets out of bed and tiptoes to the landing, where she can just hear her voice.

'She's still very upset about Erin,' her mum's saying. Who's she talking to? Dr Haslam? 'And she's not stupid – it wasn't a mistake when Oxford offered her a place, you know... She was the brightest in her year at Clapton... I wish we could tell her the truth... Of course I won't... It's just that sometimes I wonder if we're doing the right thing.'

A long pause as Haslam, if it is him, talks. The *truth*? What aren't they telling her about Erin?

'I get that,' her mum says. 'I just find it incredibly hard to lie to my own daughter. I know it's for the best, but you must understand how difficult this is for me?'

More from Dr Haslam before her mum speaks again. 'She's just working as a secretary on the lifestyle desk... I really can't see the harm... It's good for her, keeps her mind busy... Well, I disagree.' Her mum's agitated now, angry. 'She's not actually a journalist... It's an amazing opportunity. A chance to find her feet after three years at that wretched institution.'

Her mum really didn't like Oxford. After another brief exchange, she signs off, clearly troubled by their conversation. Bella is too as she slips back into her bed. *Just working as a secretary... not actually a journalist.* She'll show her mum, prove Dr Haslam wrong.

Half an hour later, Bella creeps downstairs. She can't sleep, not after what she has overheard. Her mum's lying on the sofa in the sitting room, eyes closed, a book on her lap. *Everything You Have Told Me is True: The Many Faces of Al Shabaab.* She's promised to take Bella to Somalia one day, but she thinks it's still too dangerous. They used to live in the safety of Mombasa while her dad travelled to and from Mogadishu in Somalia. Bella hovers in the doorway, wondering whether to wake her. Instead, she turns off the main light. Her mum's always got too many lights on. The darkness seems to stir her.

'Hi, sweetie,' her mum says, looking around. 'You OK?'

'Who was that on the phone?' Bella asks.

Her mum hesitates before standing up and walking barefoot into the kitchen.

'We have to give each other space, you know, if we're going to live together like this,' she says, her back to Bella at the sink. 'I'm trying to give you yours and I'd appreciate it if you gave me mine in return.'

'What did he want?' Bella asks, watching as she washes a couple of plates and stacks them on the drying rack. 'Dr Haslam?'

Her mum spins around. 'So you were listening.'

'I just need to know if Erin's alright.' She pauses. 'If she's alive. No one's telling me anything.'

Her mum takes time to answer, drying her hands on a tea towel. 'She's still not well.'

Bella sighs with relief and turns off the main kitchen light, leaving a solitary lamp on above the sink – more than enough to see.

'And will you please stop turning off all the lights?' her mum snaps, throwing down the tea towel. She turns on the main light again, clicks on the kettle and opens a cupboard above her.

'It's a waste of energy,' Bella says. 'And as you've still not converted to a green supplier—'

'I like having the lights on,' her mum interrupts, dropping a herbal teabag into a mug. She pauses and gets out another mug and teabag. 'Particularly at night. When it's dark.'

'There's still plenty of light in here,' Bella says, nodding at the lamp.

Her mum turns and leans back against the sideboard, arms folded like a schoolteacher. She's no longer cross, even though she has every right to be. Bella's pushed it and this is her mum's house, her electricity bill. Instead, she smiles and wipes away a trace of a tear.

'I love you, sweetie. Your mission to save the planet. But sometimes you need to stop and think a little about others. Particularly when you're sharing a space with someone. A home. I like having the lights on because I couldn't sleep that first night I heard your dad had been killed. There was a power cut – there were a lot in Mombasa in those days

– and the house was dark. Too dark. So I like having the lights on now. OK?'

Bella stares at the floor, flooded with guilt as she bites her lip. She remembers that night so well, the candle in their room. Why hasn't she mentioned this before?

'I'm so sorry, Mum,' she says, walking over to hug her. 'I didn't realise.'

They move back into the sitting room and settle down on the sofa with their herbal teas.

'There's a lot you don't know about me,' her mum says, a hand on Bella's knee. 'About us. Life. But that's OK. You've got an inquisitive mind. And your dad hoped more than anything that you'd both grow up curious about the world.'

'So why doesn't Dr Haslam want me to be a journalist?' Bella asks, leaning forward to take a sip of her tea.

'You really were listening to me on the phone, weren't you?'

'It's not his business what I do now anyway,' Bella continues. 'I don't understand why he's still ringing us.'

Perhaps Dr Haslam is her mum's boyfriend. The effort she'd made to look nice when she picked her up from college. Is that what her mum meant about having to lie to her own daughter? She's an attractive woman and he's a good-looking man. Not Bella's type, but her mum wouldn't be the first to fancy him. Some of his male students even started to dress like him in their final year, complete with small, round reading glasses, skinny jeans and corduroy jacket.

'Because he cares about you,' her mum says. 'And cares about Erin. He was responsible for your well-being for three years. Head of pastoral care, or whatever it's called.'

'But he's not responsible now.'

'No, he's not.' Her mum gets up from the sofa. 'I'm tired, sweetie. Time for bed.'

Bella watches as her mum climbs the stairs.

'I'm not just a secretary, by the way,' she calls up after her. 'I think I've got my first story. My first byline.'

Her mum stops and turns to look back down at her.

'That's wonderful,' she says. 'If you really believe you're a journalist, then maybe that's enough.' She pauses. 'Show Dad what you can do.'

12

Silas

Silas is not sure if it's possible for an underground community to go to ground, but that's what seems to have happened with the assorted collectives, pranksters and artists who are responsible for creating most of the 'better' crop circles in Wiltshire. The circle-makers, as they are known. He's been hitting the phones all morning and can't get hold of any of them. No big surprise. Having a dead body at the centre of a circle would not be a part of any design drawn up by these people, who might be time-wasters but are generally peace-loving, in his experience. And this morning's sensational coverage in the tabloid press seems to have sent them diving for even deeper cover.

'He hasn't been answering his phone so I doubt he'll answer the front door,' Silas says to Strover, as they drive down a long farm track in their unmarked car. Noah, the circle-maker they've come to see, is someone Silas met a few years back, when a farmer held him at gunpoint after catching him in the act of flattening one of his wheat fields at 2 a.m. Silas managed to defuse the situation – he usually

56

sided with the landowner, but on this occasion the farmer was being an idiot – and Noah has been grateful to him ever since. He has also specialised in mathematical circles in the past, coded representations of equations and formulae. Silas remembers one in particular that looked like an atomic structure: neutrons and protons and revolving electrons. Several circle-makers that Silas has managed to talk to have told him to try Noah, as the cryptic structure of the circle at Hackpen Hill bears all his customary hallmarks.

There's a car outside the run-down farmhouse but the place looks empty. Silas feels a pang of unease as he pulls up outside a converted barn overlooking the courtyard, where Noah holds photographic exhibitions. Somewhere in the distance, a dog starts to bark. Silas brought Mel out here once for a private view, just before they broke up, in an attempt to impress her, show off his more sensitive side. When he stopped off at one of his favourite pubs that just happened to be on the way back, she saw through the ruse. She was more forgiving when he came home at midnight yesterday. A quinoa salad had been left out for him, which he dutifully ate, trying not to think about the kebab he'd devoured on the way home.

Silas and Strover step out of the car and peer in through the barn window. The walls are covered with photographs and designs for crop circles. In the corner, some planks of wood or 'stomping boards', rope, poles and tape measures – the tools of Noah's trade that he'd talked about at the private view. A half-empty mug is on a draughtsman's table, pencil and ruler beside a diagram.

Silas stands back, looks out across the silent, neolithic landscape: Milk Hill, Walkers Hill, and on towards the ridge-worn flanks of Martinsell. There are no houses in sight, just the burnished plains of the Vale of Pewsey below and the Marlborough Downs to the north, rolling like a vast, swollen sea. They buried their dead here five thousand years ago, deep within the chambered tombs of long barrows that still exist today. The view can't have changed much since then – except for the white horse below Milk Hill, chalked onto the hillside in the early nineteenth century.

Noah captured the essence of this ancient landscape in his black-and-white photos, which usually featured one of his own circles. Silas glances around the outbuildings again, remembering the private view, the talk Noah gave that night. Mel described him as a gentle soul, a softly spoken artist. No doubt good at yoga too. And he seemed to have arrived at a happy compromise with the believers in the croppie world, claiming that paranormal energies acted through him while he made the circles. Inexplicable forces called him to particular fields in the night. He'd even seen balls of light in the sky that guided and inspired him while he worked. He played along, in other words. Everyone a winner. So where the hell's he gone now?

'Sir, we need an ambulance,' Strover says. She's walked over to the main house and is peering in through a lower window.

Silas rushes over to her. Cupping his hand against the glass, he can see Noah lying on the stone floor of the kitchen, dark blood pooling around his head like some sort of perverse halo. There's evidence of a recent altercation, chairs knocked over, crockery broken. The place is a mess.

'Christ, you call Control, I'll go in,' Silas says, running

around to the front door, which is locked. A second later, he smashes a lower window in the sitting room and climbs through, glass crunching under his feet on the floor. It's a while since he's had to force an entry. Crucial evidence at the scene will be disturbed but a life is at stake.

Noah is still alive, but barely breathing. His pulse is weak too. He wasn't the strongest person anyway. Silas makes him comfortable with a cushion, searching his body for injuries. He's taken a heavy blow to the back of the head but otherwise he seems unscathed.

'Noah? Can you hear me?' Silas asks, his mouth close to Noah's ear, as Strover comes into the kitchen. 'It's DI Hart. Silas Hart. Swindon CID.'

'Air ambulance on its way, sir,' Strover says, putting her radio away.

Noah opens his eyes and swallows.

'Get me some water,' Silas barks at Strover. A moment later, he puts a glass to Noah's lips and he takes a sip.

'They wanted to know who commissioned the circle,' Noah whispers.

'Who did?' Silas asks.

'I don't know who they were. They drove up here in a Range Rover. Black… tinted windows. I thought they were clients, down from London.'

'Number plate?' Silas asks, more in hope than expectation.

Noah shakes his head and then starts to speak. 'RO something?'

A Reading number plate. Silas glances over at Strover, who makes a note.

'And you didn't tell them,' Silas continues, as he puts the glass against Noah's lips again. He takes another sip, closes his eyes. Silas is losing him. He's surprised Noah didn't reveal the name, impressed with his bravery. Particularly in someone so slight. He could have been killed by the blow to the head.

'Noah, this is important,' Silas persists, louder now. Strover catches his eye with another one of her looks. He knows Noah needs to be treated with care, but this might be their only chance. 'Why didn't you tell them?'

When he finally answers, Noah's voice is faint. 'I didn't realise they were going to place a dead body in it.' He pauses. 'Of course I didn't. I wouldn't have touched the job if I'd known.'

'This person who commissioned the circle – did he ask you not to tell anyone?' Silas asks.

Noah nods.

'And you're not going to tell me who it was either?' Silas asks.

Noah shakes his head.

'This has become a murder inquiry now, Noah,' Silas continues, failing to suppress his frustration. 'It's gone way beyond making pretty patterns in a field.'

Noah closes his eyes. Silas thinks he's lost him again but then he speaks.

'I never knew his name. We didn't meet. We talked on WhatsApp, he transferred half the money up front, sent me the design. That's how these things work.'

Silas remembers a rash of above-board commercial logos that appeared in crop circles a few years back. Nike, Greenpeace, Weetabix. The *Sun* even commissioned a circle to promote its campaign to bring the Olympics to Britain. This one is different. Off the books.

'And you've deleted the conversation on WhatsApp,' Silas says.

Noah nods again. It has to be bloody WhatsApp. Messages are end-to-end encrypted, impossible to recover once deleted.

In the distance, the sound of an approaching helicopter.

'Put an obs request on the black Range Rover,' Silas says to Strover, getting up from the floor where he's been holding Noah. 'Tinted windows, RO number plate. It can't have gone far.'

'Sir, I think he wants to tell you something else,' Strover says, nodding at Noah.

Silas turns and crouches down beside him again, his ear close to Noah's lips.

'It's not the only one,' Noah whispers, his voice faltering.

'There are others?' Silas says, glancing up at Strover. 'How many? How many circles did you make?'

But this time Noah has no strength left to speak. Instead, he manages to hold up two fingers before his eyes shut and he falls limp in Silas's arms.

13

Bella

Hey sis

Can't seem to get through on FaceTime so I thought I'd send you an email. I know, I know, they're bad for the environment too – did you know that if we all sent one fewer email a day in Britain, we'd save more than 16,000 tonnes of carbon a year? That's equivalent to more than 80,000 flights from London to Madrid. Crazy! At least I'm still resisting a smartphone and using that old brick you left behind. Could become a problem at work but so far, so good. There are several girls in the office who have brick-phones too, trying to wean themselves off TikTok, so it's not just your uncool lil sis.

I might have got this wrong but I think there's something between Mum and my old tutor at Oxford, Dr Haslam. I know, don't want to go there, but Mum's been behaving so strangely, talking in whispers on the phone when he rang last night. She's been fussing over me quite a bit too, like she used to when we were young and I had a stomach bug. To be honest, I think it would do her good if she did have a relationship. Get her out of the house. And it's been so long. I'm just not sure about Dr Haslam. You'd know what I mean if you met him. And I suppose the whole thing feels a betrayal of Dad. Which is dumb, as he'd want Mum to be happy more than anything, not still moping around fifteen years after he died.

Do you remember when we came back from the hospital in Mombasa that day? I've been thinking a lot about it recently. You rushed upstairs to your room and cried and cried while I sat on our ayah's lap on that big wicker chair in the hall. And Cadogo wouldn't stop barking in the yard. Even he knew Dad was dead. It was only me who refused to accept it. I never shed a tear.

Let me know about graduation. It would be epic if you could come over – by sea, of course, preferably a sailing boat (only kidding). You know Mum would love it too.

xxx

PS Guess what? I'm checking out a pub for work called the Slaughtered Lamb. You know that was the name of the pub in *An American Werewolf In London*, which you made me watch when I was ten. Ten, Helen. Stay off the moor…!

14

Silas

It's been a while since Silas has seen Wiltshire from the air and he's forgotten how beautiful the county looks in the low, late afternoon sun. Up ahead, Fyfield Down, harsh and ancient, littered with the same sarsen stones used at Stonehenge and Avebury. Beyond it, the Vale of Pewsey. And behind him, the sprawling conurbation of Swindon. Even his home town doesn't look so bad from up here, glistening in the dusk.

They'd passed over the Magic Roundabout shortly after the police helicopter picked him and Strover up from behind Gablecross and the layout of the town's famous ring junction had suddenly struck Silas as some modernist take on a crop circle – five satellite roundabouts around a central, larger one. All this talk of UFOs and plasma vortices is beginning to get to him.

'Where do you want to start, boss?' the pilot asks Silas over the radio. As part of the National Police Air Service, the helicopter had to come out from RAF Benson in Oxfordshire. Long gone are the days when Wiltshire Police had its own chopper. The pilot is normally accompanied by two tactical flight officers, who feed back live TV footage to the Control Room, but Silas has persuaded NPAS to let Silas and Strover take their places. It isn't orthodox, but nor is this particular search mission. It also saves on costs.

'Hackpen Hill, north-west of Marlborough, and then work southwards,' Silas says.

Noah was taken to hospital before being able to provide them with any further details, which didn't go down well with DCS Ward. Did Noah's raised fingers mean two more circles? Or two in total? Hope can be a dangerous thing, but Silas is working on the assumption that they're looking for one other circle.

'Do you have any idea how much that chopper costs an hour?' Ward had asked, when Silas had gone to see him with his begging bowl.

'We have good reason to believe that when we find the second circle, we'll find another body,' Silas had said.

'Three grand an hour. I could do a lot with that.'

Silas had ignored Ward, pressed on regardless. 'And given that no one has contacted us yet about the discovery of a second body, we can safely assume the second crop circle is only visible from the air.'

'What would be the point of that?' Ward had asked.

'Some of these circles aren't discovered for days,' Silas had replied, pausing. 'All adds to the mystery, apparently. Their psychic power.'

Ward had raised his eyebrows at that, expecting a smirk in return. But Mel had given Silas a hard time over breakfast, told him not to be so tough on croppies, keep more of an open mind in life, so he had looked away from Ward. Mel would have been proud of him.

'Others are done in conjunction with drone photographers,' Silas had continued. 'Who duly flog the photos to the tabloids.'

'But not this one.'

'No, sir. This one's different. And we need to find it quickly. Before the media.'

Silas didn't have to spell out the practicalities of a human body lying for days in the countryside in high summer, even if it had been deep-frozen first. Animals could be the biggest problem, destroying vital evidence.

Back in the police helicopter, Silas spots three crop circles in the first twenty minutes, using his own Swarovski bird-watching binoculars. A fiftieth birthday present from Mel and better than anything work might provide. One crop circle is outside Stanton St Bernard in the Vale of Pewsey, another over by Etchilhampton Hill near Devizes and a third one close to Bratton Castle, on the western edge of Salisbury Plain. All three impressive in their own way, but none of them complex, like the one at Hackpen Hill.

'You sure that's not done by aliens?' the pilot asks, as they hover over the circle at Bratton Castle, across from a white horse on the hillside.

'I'm not saying anything,' Silas says, taking in the rich geometric pattern, a six-pointed star with jagged, triangular edges.

'It's a fractal formation,' Strover says. 'A "Koch snowflake", named after Helge von Koch, a Swedish mathematician.'

Silas looks up at Strover and shakes his head. She never ceases to surprise him. 'More bedtime reading?' he asks.

'Went down a bit of a rabbit hole online,' she says, gazing

at the circle. 'If you look closely, each smaller pattern is the same as the whole.' She pauses. 'Quite beautiful, actually.'

'But not what we're looking for,' Silas says.

Ten minutes later, as they fly low over the hamlet of Oxenwood, Silas thinks he's seen something. He asks Strover for the binoculars.

'Over there,' he says to the pilot, pointing in the direction of Andover in the far distance. They are close to the Wiltshire/Hampshire border.

'Got it,' the pilot says. Silas's stomach lurches as the helicopter drops altitude. He shouldn't have had that kebab last night.

It takes a second for him to get his bearings. There's a pub near here, at Tangley, where he used to cycle when he was fitter. Lighter.

'That's Chute Causeway,' the pilot says, pointing at the map on the screen in front of him. 'An old Roman Road.'

The spot they're heading for is two fields back from the road and, unlike Hackpen Hill, there is no surrounding higher ground from which to observe the pattern. It can only be seen from the air, just as Silas thought, and the area is remote.

Silas falls silent. He's got the binoculars trained on the circle, another complex pattern that once again involves hexagons, as well as an array of other geometric symbols and a large adjacent spiral disc. But he's not looking at the shapes. He's studying the human body at the centre of one of the hexagons. What's it wearing? And why doesn't it appear to have any arms?

There's no tactical flight offficer to operate the onboard cameras so Strover pulls out her phone and starts to take pictures of the scene below. Silas watches, transfixed. A flock of birds lifts up from the body as the helicopter moves in to land beside the circle. Corvids are often hard to distinguish, particularly in flight, but Silas knows at once what these are. Black, oily plumage, bare white face.

Rooks.

15

Bella

The Slaughtered Lamb is busier than Bella expected and the conversation at the bar is about a body that's been found in a nearby crop circle. She'd read the evening newspaper report on the train but hadn't realised the circle was so close to the pub.

'Has anyone worked out what the circle means yet?' one of the locals asks. Bella clocks an Irish accent, thinks of Erin. She needs to keep it together.

'Here we go,' the landlord says under his breath. 'Another one of Sean's famous conspiracy theories.'

'Maybe it's a sacrifice of some sort – a ritual gone wrong,' a second local suggests, winking at the barman. '*Wicker Man* comes to Wiltshire.'

Bella heads over to the pub's only corner table, ears straining to catch every word. *The Wicker Man* was another film Helen made her watch when she was too young. She's still haunted by the clifftop finale. There's a sign on the table saying, 'Reserved, 8 p.m.' She'll take her chances. Sitting down, she starts to write in a notebook, hands tingling with anticipation. She prefers pen and paper to using a phone. It's the same with newspapers. She grew up with them scattered around the house like laundry and she likes to hold one in her hand rather than read the news online. Old school all the way.

'It's definitely a formula of some sort,' Sean says. 'Which means there's only one suspect.' He pauses for dramatic effect, milking the moment. 'Porton Down.'

Bella doesn't know a lot about Porton Down except that it's a secretive government site in Wiltshire. She glances around the pub. This is what she dreamt of. Her dad would be so proud. *Just don't go mad with the bar bill.* She takes a sip of her tomato juice and smiles to herself, remembering Mark's words when she asked if she could have a go at the 'Overheard' column.

The pub door swings open and a young man wearing thick black-rimmed glasses walks in.

'Palmer will know,' one of the locals says, making room for the new arrival at the bar. 'Won't you, Palmer?'

'Know what?' the man says tersely. He is tall and broad-shouldered – a rower's frame. Bella never did get to try that at uni. His young face is smooth, his gait a touch clumsy. Maybe it's his feet, which are big and point outwards, like a penguin's. He glances around the pub, notices Bella in the corner and turns away, pushing his glasses back with one finger. Shy too.

'You must know what that coded crop circle means in the field where they found the dead body,' Sean says. 'Got to be something to do with your lot. Got to be.'

Bella watches as the barman serves Palmer. He's about to leave the bar with his drink when another man, further down, leans over in his direction and addresses Palmer in a thick Wiltshire accent:

'Come on, Palmer, it was more than a coincidence that the first ever nerve agent attack on UK soil happened just down the road from Porton Down, where novichok's been kept for years.'

The tone is confrontational, almost angry, and the whole bar falls silent.

'Are you asking or telling me?' Palmer says, taking a sip of his drink. He seems nervous, eyes darting around the room.

'I'm asking you to put our minds at rest, given you work there. Tell us what's really going on. I'll bet my house the corpse in that field's another bloody Russian agent taken out by Putin.'

'Leave the poor man alone,' the landlord says, winking at Sean. 'One conspiracy theorist in my pub is quite enough.'

Bella reaches down for her bag as Palmer comes across to her corner, stepping over a couple of sleeping black Labradors.

'I'm sorry, is this your table?' she asks, getting up from her chair.

'It's OK, please,' he says. 'In fact, you'd be doing me a favour if you stayed.' He nods at the bar, where the locals have resumed their discussion. 'Might give them something else to talk about for a change.'

One of the locals looks over in their direction. Bella stares him down and the man turns away. She's still standing, torn between leaving and talking to this man who appears to work at Porton Down. The same man she was told to 'overhear' by the anonymous letter writer. No self-respecting journalist would walk away from such an opportunity. He also happens to be quite cute, in an awkward sort of way. Nicely dressed in a crisp white shirt and blue jeans. And elegant, long fingers.

'What you having?' he asks, gesturing at her half-empty glass. They are both still standing.

'I'm fine, honestly,' she says, noticing a scar on his forehead.

'Come on. Bloody Mary, is it?' he asks.

'Just a tomato juice.'

'On the hard stuff,' he says, spilling some of his drink on the table as he raises his glass in solidarity. 'Me too.'

Bella watches as he mops up the spillage with a paper napkin. He makes a mess of it, knocking his glass again.

'Anyone would think I've been drinking,' he says, laughing. 'It's only ginger beer. Honestly.'

Bella smiles. Why's he so nervous? 'Let me get mine,' she says. 'Save you having to run the gauntlet again.'

He looks over at the bar. 'OK, but put it on my tab. I insist. Jim Matthews.'

Bella looks surprised. 'Not Palmer, then?'

'No,' he smirks. 'Palmer's just a nickname. You know, because of the glasses.' He looks down, embarrassed. Bella stares blankly at him. She has no idea what he's talking about. 'Harry Palmer?' he continues. '*The Ipcress File*? Michael Caine?'

'Right,' she says, bluffing. 'I'm Bella.'

'Nice to meet you, Bella,' he says, shaking her hand. *And very nice to meet you too.* 'You're not a cop or a spook, are you?' he adds, finally sitting down.

'No,' she says, taken aback by the sudden change of tone.

But she is a journalist. Her face blushes. Should she sit down too or go to the bar?

'Not that you'd tell me if you were, of course,' he laughs, unaware that Bella's stomach has just flipped.

'Why would I be?' she asks. She blushes again, certain that her embarrassment must be obvious.

'Someone followed me back from work tonight,' he says. 'That's all.'

'From Porton Down?'

He looks up, fixing her in the eye for a second, and nods. Why's he telling her this?

'Maybe it was the Russians,' she suggests, trying to disguise her interest as a joke.

'It wouldn't have been them,' he says, taking her comment way too seriously. 'They want me to live. This person tried to drive me off the road.'

16

Silas

Silas walks across the field with a heavy heart and his head ducked low as the blades of the helicopter spin down. He'd hoped there wouldn't be another body, but he'd feared they'd find one as soon as Noah told him about a second crop circle.

The victim, a young woman, is lying on her side, but it's the clothing that draws Silas's eye. She is trussed up in a white canvas straitjacket, secured at the back by five webbing straps and buckles. The arms are crossed at the front of the body and secured with a further vertical strap to the collar. No wonder he couldn't see her arms from the air. Another strap passes under the crotch. Her legs are bare, covered in dust and mud.

'Jesus,' Silas says, looking at the woman's bloated face. The rooks have been busy, pecking at her eyes. He shoos away a fly. At first glance, the scene has all the hallmarks of a sexual crime.

'What do you think?' Silas asks Strover, squatting down to take a closer look at the jacket.

'Poor woman,' Strover says, shaking her head.

Silas reaches out and touches the victim's cheek with the back of his hand.

'Cold,' he says, turning to look at the muddied fabric of the jacket. What exactly's happened here? In his experience – professional, not personal – straitjackets used for sexual gratification tend to be black and made of faux leather. This is definitely tough white canvas.

Strover squats down beside him.

'The medical profession stopped putting people in these years ago,' Silas says, nodding at the jacket. 'Magicians still use them – I once saw a video of Houdini escaping from one, suspended upside down from a crane. And they're popular in the BDSM community.'

They both stare at the body in silence. 'This isn't BDSM,' Strover says. 'There's no O-ring – on the collar.'

Silas turns to Strover, puzzled. 'O-ring?'

'For the leash,' she says, standing up.

Silas stands up too, feeling slightly queasy. She's right. This is the real deal, the stuff of asylums, a crude restraint designed to prevent psychiatric patients from harming themselves or others. A legacy of a bygone era in psychiatry.

'The first victim was lobotomised and the second is trussed up in a medical straitjacket,' Silas says. 'They're both—'

'Barbaric,' Strover interrupts.

Silas glances up at his colleague as she turns away. In the two years they've worked together, she's proved herself remarkably resilient, untroubled by the various homicides they've investigated. But something about this case has gone to the core.

'There's also another code to crack,' Silas says, looking around at the flattened wheat as Strover walks away. 'You OK?' he asks.

She raises one hand, just needs a moment.

Silas has learnt when to give Strover space. He turns back to the body, thinking of the earlier victim. Each crime scene appears to have been staged, freighted with meaning, but it's not an obvious case of serial killings, of a real-time murder spree. He suspects that this body, like the first one, came from a morgue and has been long dead. Bending down, Silas touches the back of his hand against her cheek again. Definitely too cold for a corpse that's been lying in a field in summer.

'Whoever commissioned Noah to make these circles is also responsible for the bodies,' Silas says.

'For killing them too?' Strover asks.

Silas looks up, intrigued by her tone of doubt.

'Presumably,' he says, pausing, but he doesn't want to shut her down. 'Maybe not?'

Strover takes her cue. 'What if the people who placed the body here weren't responsible for her death?' she asks, walking back over to Silas. 'The way both bodies have been positioned, they seem to be part of the overall design of the crop circles. The vision. And if the lobotomy was carried out just before the circle was made – that would seem to be part of the same process too. Ditto the straitjacket.'

'Unlike their earlier deaths, you mean,' Silas says. According to Malcolm, the first victim died a month ago.

'I just think we should consider that whoever commissioned the circles, and placed the victims in them, might be trying to draw attention to the manner in which they originally died, that's all,' Strover says.

It's a plausible theory. And he's pleased that she feels confident enough to put it forward, to challenge Silas's interpretation of events. It's what makes them a good team. The coded patterns of the crop circles, the lobotomy and the straitjacket – they're all signs, not causes, of death, part of the same cryptic message. More semiotics, as Strover might say. Establishing where the bodies came from should help to decode that message.

They're still checking the first victim against missing person records, which might throw up a match. Maybe this woman has been missing too? Hundreds of people disappear each year, some more missed than others. Conor vanished for a while, the worst six weeks of his and Mel's lives. If this woman's disappearance wasn't reported, there's little chance of identifying her.

'Did you get a good photo of the pattern from the helicopter?' Silas asks.

Strover nods. 'I've already sent it over to Cambridge.' She doesn't hang around. 'And to Imperial, the chemistry professor. We're talking in the morning.'

'Tell them the body was found in a straitjacket. It's important your boffins know all the details. There must be a connection between the symbols and what's been done to the victims. And get CSI down here now.'

Silas looks around at the patterns again, the hedgerow in the distance, sniffing the summer air for clues. Dizzy midges are dancing in the evening sunlight. A solitary buzzard twists and turns above them in the warm thermals. Someone would have had to park up on Chute Causeway and carry the body through two fields – maybe drag it, given the dirt on the victim's bare limbs. He can't see any other way to approach the field.

He walks back over to the body. She too looks as if she's fallen from space and is now hugging the earth's surface for comfort, her cheek pressed against the ground, hands clutching at the flattened wheat. And then he realises that it's not dirt that's covering her arms but tattoos. He bends down and studies them more closely. A fine pattern of dark feathers.

17

Bella

'You're not tempted by the local beers, then?' Jim asks, as Bella returns from the bar with her tomato juice and sits down opposite him on a low stool, not sure what to do with her legs. Helen used to get her to cross them twice, a trick Bella could do with worrying ease.

'I don't really drink, to be honest,' she says, tucking them under her as best she can. 'My nickname was Gerald at school,' she adds, noticing that Jim's clocked the manoeuvre.

This time it's his turn to stare blankly at her. 'The tall giraffe in *Peppa Pig*?' she asks, but he shakes his head and they both laugh.

She wonders how much to tell this man whom she's known for barely five minutes. But something about him, his open face and honest eyes, makes her want to say more than she should. She also feels a certain unspoken kinship with him. By her reckoning, they're the only two in the pub who are under fifty. Palmer and Gerald. The odd couple.

'I've just finished at uni,' she says.

'Drying out, then,' he says, arpeggioing his fingers on his glass. 'Where were you?'

Bella hesitates, watching his fingers until they stop. She was once told that someone reveals they've been to Oxbridge within fifteen minutes of meeting them. She's determined not to be that sort of person. Her time at Oxford is her secret, not a source of bragging rights.

She takes a sip of her tomato juice.

'Let me guess,' Jim continues, sensing her hesitation, drumming his fingers again. 'English at Oxford.'

Bella almost chokes on her drink. Is she that easy to read?

'Am I right?' he asks, slapping his thigh as Bella puts down her glass. 'I am right, aren't I? I knew it.'

'And you?' she asks, but before she has time to guess, he looks at his phone buzzing on the table. His smile disappears.

'I got a friend to check out the number plate – of the car that tried to drive me off the road,' he says, reading a text. His left leg starts to bounce beneath the table. 'Black Range Rover. Tinted windows. Asked him to find out who it belongs to.'

'And?'

Jim looks up as someone walks into the pub. Bella looks too. She's no longer as relaxed as she was. Her head is swimming with images, of cars crashing in the night. Headlights. Headlines.

'Going back to your choice of drink,' he says, pointedly changing the subject. 'I may be wrong, but I don't think you're the type who spent three years getting off their head at uni.'

If only he knew. Her first year was the worst. 'We haven't

even established where you were yet,' Bella says. Why doesn't he want to tell her about his friend's text? Who owns the Range Rover? She decides to play for time.

'OK, fair enough,' he says. 'Your go.'

Jim sits back, eyes flicking across at the locals, who seem to have lost interest in him.

'I'm guessing you studied a science, given you work at Porton Down,' Bella begins. 'Chemistry at Cambridge?'

'Not bad. The chemistry bit's right.'

'Imperial?'

Jim shakes his head.

'Warwick?' she asks.

'Bingo.'

'And you've just graduated?'

'Why do you say that?' he asks.

He's obviously sensitive about how young he looks.

'You've recently finished a PhD,' she adds, hoping to flatter him by adding a few more years.

'I graduated four years ago.'

'OK.' He really does look young for his age.

'And you've been at you-know-where ever since?' she asks, smiling.

'A three-year secondment – or "deployment" – to an affiliated facility at Harwell Science and Innovation Campus, but basically yes. I'm a government scientist – high-functioning, mind – and the Defence Science and Technology Laboratory at Porton Down – we call it "The Lab" – is my life. For my sins. I am allowed to tell everyone that much. It's all on LinkedIn anyway – apart from Harwell. That's a bit more...'

'Hush-hush?' Bella offers.

'Exactly. Hush-hush.'

Jim breaks into a smile at the sound of the words, as if he's trying on a suit for the first time and likes the cut of it.

'And do you enjoy it?' she asks.

'Enjoy? It's a pain to get to – Porton's in the middle of nowhere – but they lay on lots of social stuff for employees. And we breed like lab rabbits as a result. There must be at least three hundred couples on site – out of a workforce of more than three thousand. Quite the family business. And I reckon at least 20 per cent of us are on the spectrum. The Lab's very enlightened like that – big on neurodiversity.' He pauses, grinning. 'If I tell you anything else I'll have to shoot you. And I don't want to do that. I'm rather enjoying our chat.'

'Me too.'

They look at each other, smiling awkwardly. And then she's overcome with a sudden need to share with this stranger the real reason she's off the booze.

'I'm not drinking because...' She pauses. 'I'm working,' she says, feeling uncomfortable that she hasn't already put her cards on the table, revealed to Jim that she's a journalist.

'Nice work if you can get it,' he says. 'Sitting in the corner of a country pub on a Friday night. Who do you work for? *The Good Pub Guide*?'

She takes a deep breath. Maybe she should tell him?

'I'm a journalist,' she says. 'For a national newspaper.'

Jim pauses, drink to his lips, as if he's just tasted poison. After what seems like an age, he puts down his glass.

'I should have known,' he says.

It's like his body's suffered a puncture, collapsed in on itself. He shifts in his seat, half turning away from her, his eyes looking upwards, indicating that he's through with their chat.

'It's probably best you go now. I could lose my job,' he adds, checking his phone again. 'It was nothing serious on the way home – just a bit of careless driving on my part.' His tone is cold, indifferent. 'I've been very tired recently. It's been nice meeting you.'

The conversation is clearly over. Bella reaches down for her bag and stands up, caught off guard by a surprising wave of emotion. Her love life at Oxford was a non-starter, no memorable romances, no connections that meant as much as her friendship with Erin, but there's something about Jim that's touched her. The mix of confidence and vulnerability, his goofy charm and obvious intelligence. Those flattering eyes that seem to widen with wonder when he looks at her. Maybe she's deluding herself and it's just the heavy prescription on his glasses.

'Nice meeting you too,' she says, struggling to keep it together. She aims for the door, head down, walking fast.

'Palmer's date went well,' she overhears one of the locals say.

She won't be putting that in the column.

18

Silas

Silas directs the CSI manager up from the road, after briefing him about the straitjacket. There's been no progress deciphering what the crop circle patterns might mean, despite numerous theories put forward by armchair codebreakers on the web. And no reports from the country's mortuaries about missing cadavers. Silas's focus now is trying to establish who commissioned Noah to make the circles and placed bodies at their centre.

'Any word from the hospital?' Silas asks Strover, who's just come off the phone. Roadblocks are now in position at both ends of Chute Causeway and Air Traffic Control has been asked to keep the area clear. A drone hired by a tabloid newspaper has already managed to take pictures of the first crime scene at Hackpen Hill.

'Noah's barely conscious,' Strover says. 'The hospital's not keen on us visiting tonight.'

'They never are,' Silas says, stepping aside to let a team of CSIs walk up through the field. A tent has been erected to cover the second body. 'We'll need to talk to him.'

What worries Silas is that Noah's attacker might reach the person who commissioned the crop circles before the police do. Why are they so keen to find them? To stop any further coded circles from appearing? CSI has already got tread marks from the dusty drive that leads down to Noah's farmhouse but there have been no sightings of any tinted-window Range Rovers with number plates beginning 'RO' in the area. Strover has also analysed Noah's recent bank transactions but nothing has shown up. No doubt the payment was made using untraceable bitcoins. Another question for Noah.

Silas is about to walk back over to the crime scene when a familiar car pulls up on the road. The driver gets out and opens the rear passenger door. It's his boss, DCS Ward.

'What the hell's he doing here?' Silas whispers to Strover.

'Thought I'd come and take a look myself,' Ward says, walking through a gate into the field.

'Good idea, sir,' Silas says, stepping aside.

Silas has learnt over the years that everything his boss says and does is best viewed through the prism of ambition, how it might advance Ward's own career. This is an unusual case, one that's now attracting international media attention, but it wouldn't normally warrant a personal appearance by the boss.

Silas escorts Ward up to the crime scene, where they both don white oversuits and step into the tent, leaving Strover to make more calls. The air inside is hot and stuffy, even though the sun has now set. Arc lights illuminate the body, still lying on the flattened wheat in the straitjacket.

'There's been a lot of idle gossip online that Porton Down

might in some way be responsible,' Ward says, glancing around the tent, which is empty apart from them.

'So I gather,' Silas says. 'Along with UFOs and time travellers.'

'As I'm sure you know,' Ward continues, in no mood to joke, 'we have a very delicate relationship with Porton. They've long memories over there and haven't forgiven us for launching an investigation into the death of a twenty-year-old volunteer at Porton – almost fifty years after he died in a sarin experiment.'

Silas remembers the investigation well, the commotion it caused in Westminster. Wiltshire Police's Operation Antler dug up all sorts of stuff that led to a second inquest into the death and a sizeable MOD payout to the victim's family.

'I'm in the process of rebuilding our relationship with Porton,' Ward continues, looking around the tent again. 'I wouldn't want anything we're doing here to jeopardise that. Do I make myself clear?'

'Very clear, sir,' Silas says, trying to conceal his surprise. 'We're currently not investigating any links with Porton Down and if we do—'

'You'll let me know first. I'm glad we understand each other.'

Ward nods at him and ducks out of the tent. Jesus. This case is already complicated enough without Ward throwing his weight around.

19

Bella

Bella walks down the village high street towards the train station, eyes welling as she reflects on how quickly it all went wrong with Jim. Should she have kept quiet and not told him that she was working for a newspaper? There was something irresistible about the story he told her, of a government scientist nearly being driven off the road on his way back from a secret research facility. An intoxicating whiff of conspiracy. And something irresistible about him too that stopped her from lying. She's always fancied tall men, for obvious reasons, and Jim certainly ticked that box. And he was surprisingly dapper for a mad scientist. She likes a man who takes care of himself, keeps his nails trimmed and glasses clean. He seemed kind too. Even when she dropped the bombshell about her being a journalist, he remained polite and was almost embarrassed by his own reaction. Best of all, he didn't seem to care what others thought of him. Happy in his awkward skin. She can relate to that.

She turns onto the station platform, glancing up at the

information board. The next train back to London has been cancelled because of a lack of crew. Typical. And the one after. No wonder people are still driving around in polluting cars. She checks the timetable poster beside her, trying to ignore a twinge of panic, and then her watch. Almost 9 p.m. There are no more trains tonight. The pub has rooms. She could see if they've got one available. Except that Jim will still be in the bar. Tricky.

She walks back up the high street, hoping that Mark will pay accommodation expenses. A solitary car drives past and turns into the pub up ahead. A black Range Rover. Just like the one Jim mentioned. His paranoia is infectious. There must be lots of Range Rovers in a rural county like Wiltshire.

She stops at the edge of the pub car park and calls her mum, telling her that she won't be back tonight. Things have been a bit tense at home, and her mum doesn't seem too bothered. Maybe Dr Haslam is in town. At breakfast this morning, Bella had nearly asked about him, whether they were having a relationship, but it felt too embarrassing. Helen hasn't replied to her email yet.

'What's the name of the pub again?' her mum asks, suddenly sounding more concerned.

'Mum, I'm OK.' She glances up at the hand-painted, bloodthirsty sign. 'The Slaughtered Lamb.'

'Doesn't sound very OK to me.'

'I'm on a proper story. It could be my big break.'

'Sweetie, that's great. I thought you were just writing a little lifestyle column.'

Bella rolls her eyes. 'It was, but now it's become something more serious.'

'And they want you to write it?'

'Sure. Why not? See you tomorrow, Mum.' She hangs up before she says something she might regret.

What's happened to the Range Rover that drove into the car park? No one has appeared. She strolls around to the back of the pub and spots the vehicle in the corner. It's dusk but she can just make out the profile of a man at the wheel. She glances at the windows – tinted – and turns to walk around to the front of the pub. Her hunch was right. Taking a deep breath, she steps back inside.

The group of locals has dispersed to a big window table and more people have arrived at the bar, chatting about the crop circle. Jim's nowhere to be seen but his glass is still on the corner table, half finished, a jacket on the back of the chair. She feels pleased he's still here. After booking the last room with the landlord, she stands at the bar with another tomato juice, keeping one eye on Jim's table. A moment later, he's at her other shoulder, having come out of the Gents behind her.

'Back again?' he asks, surprising her.

'My train was cancelled,' she says, blushing.

'I'm sorry,' he says.

'No worries.'

'I mean about earlier.' He adjusts his glasses, pushing them back on his nose. 'I may have overreacted when you said you were a journalist.'

'It's fine, honestly. Goes with the territory. And you're in a sensitive government job. I just wanted to be honest, that's all.'

She's happy they're talking again. Despite his cumbrous social manner, she feels a flutter of excitement in her stomach and follows him over to his table as if they had never fallen out. One of the locals looks across at them and Bella gives him a discreet finger. Erin would be so proud of her.

20

Silas

Silas squats down beside the body of the woman, shaking his head. If only murder victims could speak. Why the straitjacket? He stands up, contemplating a smoke, but the boss is still outside the CSI tent, talking to a young uniform whose dad he knew. Is Silas missing something? Should he be investigating a link with Porton Down? To date, it's all been wild speculation, the stuff of conspiracy theories.

Strover puts her head inside the tent, glancing at the body. She's still upset, her eyes red. The whole team is.

'Sir, we've got a possible Range Rover sighting,' she says.

'Tell me,' Silas says, standing up.

'A patrol car on the A4 outside Hungerford saw a black one, RO number plate. They gave pursuit but were called off on another job. They're not certain, but they think the car might have turned into the Slaughtered Lamb.'

Silas's eyes light up. It's a pub he knows well, not just for the decent pint it serves but because of several past cases. The landlord's a personal friend too.

'How long ago?' he asks, already picturing a glistening glass of Ramsbury Gold waiting for him on the bar.

'Ten minutes?' Strover offers, looking at her watch. 'And it had tinted windows.'

Sounds promising, although Silas no longer views tinted windows with quite so much suspicion, now that Mel's got them in her car. They're a must for florists – it keeps the flowers fresher, apparently. All those times he's seen drug dealers driving around east Swindon in Beamers with darkened windows... Florists, the lot of them.

'We need to get over there now,' he says.

21

Bella

'Actually, I came back to tell you something,' Bella says, sitting down opposite Jim in the corner of the pub.

'I got it, just now, in the loo,' he says. 'Saw it in the mirror.'

'Saw what?'

For the second time tonight, she has no idea what Jim's talking about.

'I had a peanut or something stuck,' he explains, rubbing his tongue against his clean white teeth as if he's doing his morning ablutions.

'Oh,' she says, smiling to herself. 'Can't say I noticed.'

Jim's not like anyone she's met before.

'That wasn't it, then?' he asks. 'What you came back to tell me?'

She shakes her head, intrigued by this man, his appealing mix of social clumsiness and confidence.

'You know that Range Rover you said tried to drive you off the road,' she begins.

His face freezes. 'What about it?' he asks.

'I think it's out in the pub car park, round the back. And there's a man sitting inside it. Thought you should know.'

Jim bows his head, forcing himself to breathe calmly. Bella's done that herself, when she wants to make a problem go away.

'I knew they'd be back,' he whispers.

'Did they really try to drive you off the road?' Bella asks. Jim leans forward.

'I'm exhausted at the moment,' he says. 'Extremely stressed at work, to be honest. I joined The Lab to do real science. It hasn't quite worked out like that. I saw the car behind me as I was driving back through Tidworth. It had been on my tail for a while. I tried to accelerate away, put some distance between us, but this car accelerated too. I nearly lost control around a tight bend in Collingbourne.'

'What about the number plate?' Bella asks, glancing at his phone on the table.

'According to a colleague,' he says, looking at a text, 'it used to be registered to the Home Office.'

'Probably a coincidence,' Bella says, playing devil's advocate. It strikes her as anything but.

'I doubt it.' Jim throws a furtive glance around the pub. 'There's an understandable culture of secrecy at The Lab,' he continues. 'We're all potential targets for the Russians, who are desperate to recruit people on the inside. MI5 keeps an eye on every Lab employee – and MI5's answerable to the Home Office. I'm not betraying any confidences by telling you that – even if you are a journalist. It's common knowledge.' Jim makes two quotation marks in the air. 'Open source.'

'What if I'm a Russian agent, posing as a journalist in

your local pub?' she asks, risking a smile. 'Here to spread vicious rumours about crop circles and novichok?'

Jim stares at her, as if he's considering whether she might really be a Russian agent. 'I haven't told you anything you wouldn't already know,' he says, taking a sip of his drink.

'But the Home Office – or MI5 – wouldn't be too happy if they walked in now and found us chatting together,' she says.

'Probably not. As I say, I signed up for the science, not the politics. Have you got a card?'

'I'm out, sorry,' she says, patting her pockets.

'I'm probably making the biggest mistake of my life, but something about your manner makes me want to trust you,' he says, his eyes widening.

'That's nice of you,' Bella says, blushing. She wants to trust Jim too, likes his vitality. She's pleased she was honest with him about being a journalist.

'And I've got a story that I want to share – that the public needs to hear,' Jim continues. 'About Porton Down.' Bella's stomach flips again. 'But there are people who want to stop me.' He looks around the pub, his eyes lingering on the door. 'For your safety and mine, I'm going to walk out of that door now and head back to my house. I always eat here – usually the game pie, cod and chips on a Friday – but I'll have something at home tonight.'

'OK,' she says, wondering why he's not more overweight. 'And is that far? Home?'

'Just across the road. Unlike Harry Palmer, I'm not a gourmet cook.'

'Here's my mobile number,' Bella says, writing it down on a sheet of her notebook. 'Any clues to what this story of yours might be about?' she asks, tearing out the page and handing it to him. 'It's not connected to these crop circle killings, is it?' She nods in the direction of the bar. 'Just that some people are saying they're linked to Porton Down.'

He smirks, looking across at the locals. 'People around here think everything's linked to Porton Down,' he says. 'Alien invasion? Porton. Novichok attack? Must be Porton.' And then his expression becomes more serious and his eyes narrow. 'But in this case... they're right.'

'How do you mean?' she asks, reassessing the locals she'd dismissed as drunken conspiracy theorists.

Jim gets up to leave, taking his jacket off the back of the chair. 'Those crop circles, they're not just fancy patterns,' he says.

'What are you saying?' she asks, wishing he wasn't being so cryptic.

He leans down, his sweet breath warm against her ear. 'Thank you for warning me about the Range Rover. I'll explain everything soon. I promise.'

22

Silas

Silas turns into the car park at the Slaughtered Lamb and slides his car into an empty space in the corner. No black Range Rovers in sight, with or without tinted windows. He opens the car door and inhales the sweet summer evening air. It's strange being back in this village. Two different cases have brought him here – and one attractive woman, the local GP, when he and Mel were separated. It's also where he found Conor, after he'd gone missing, crouched in a barn in the woods on the far side of the canal. He shudders at the memory as he walks into the pub. Strover stays outside.

The landlord seems pleased to see Silas and after a quick chat, he rings the bell.

'Anyone seen a black Range Rover in the village this evening, number plate beginning RO?' he calls out, glancing at Silas for confirmation. Silas nods.

A few dissenting murmurs about wealthy weekenders, but no one seems to have spotted the Range Rover. Silas watches with envy as the landlord pours someone else a pint. He must resist, be good. There's work to be done.

'Any word on the body in this crop circle then?' a local at the bar asks him. Silas recognises the man and his Irish accent from a previous visit. Sean, the resident conspiracy theorist and screenwriter. For a few brief weeks, he was dating Strover. Maybe that's why she was happy to stay outside.

'Not yet,' Silas says.

'Look no further than Porton Down,' he says, taking a deep draught of his Guinness. 'The UK's very own Area 51.'

'Is that right?' Silas says, glancing at his watch. He knows it's good to keep the locals onside but he hasn't got time for a chat about Porton Down. Sean will be telling him next that they keep aliens there.

Back outside, Silas is about to get into the car when he hears a noise from a house across the street. Strover clocks it too. A short, sharp shout of pain, as if someone's hit their thumb with a hammer, followed by the sound of furniture being upended. They both pause. Silas must press on with the investigation. The last thing he needs is to be caught up in a domestic; but he can't walk away, pretend not to have heard anything.

23

Bella

Bella stares at the ceiling of her room in the pub. She can't sleep, not after her encounter with Jim. Can't read her book, either. Why didn't he tell her more about the link between the crop circles and Porton Down? Does he not trust her? She thought he might say more after she'd warned him about the Range Rover in the car park. Did the driver really try to knock him off the road?

Earlier, she'd looked Jim up on LinkedIn, where it says he's been employed as a chemical analyst with the Defence Science and Technology Laboratory – The Lab – for the past four years. His degree in chemistry at Warwick is also included, but there's no mention of any secondment to Harwell Science and Innovation Campus. It sounds like his sort of place, though. Once home to the Atomic Energy Research Establishment, it discovered one of the world's highest prime numbers in the 1990s. And it currently houses the UK's national 'synchrotron', the Diamond Light Source building that scientists use to study anything from viruses and vaccines to jet engines and fossils.

Bella's phone rings beside her bed. She doesn't recognise the number.

'It's me,' Jim says. His voice is familiar but quiet, as if he doesn't want to be overheard.

'Are you OK?' Bella asks, relieved that he's finally rung her. He hadn't given her his number earlier in return.

'Not really.'

She sits up, her chest tightening. 'What's wrong?'

They should be having this conversation face to face, not on the phone.

'I'm upstairs, in my bedroom,' Jim says. 'I was about to go to sleep when I heard someone moving around downstairs.'

Is he drunk? Maybe there was vodka in his ginger beer and he was drinking a Moscow mule, Helen's favourite cocktail. He sounds sober, but what does she know? She's only just met him.

'Do you live alone?' she asks.

'Apart from my cat, who's with me on the bed. She never comes upstairs but must have heard something. Cats sense danger earlier than humans.'

She can imagine Jim with a cat, stroking it with his long fingers.

'Is the Range Rover still there?' His voice is a whisper now. 'In the car park?'

'I'll check,' she says, getting out of bed. 'It was there a minute ago.'

Bella blanches as she looks down onto the empty space.

'It's gone,' she says. 'Must have just left.'

Jim doesn't say anything for a while. 'If they take me away, you've got to write about this. Not me, but the truth.'

'What truth?' Bella asks, her heart racing now. 'You haven't told me.'

He doesn't answer immediately. 'It wasn't a coincidence you turned up in the pub today. The same day that the crop circle appeared. Someone meant us to meet.'

'Who?' Bella asks, thinking again about the typed letter, the PS urging her to talk to the man in the corner of the pub. The lurid reports of the dead body found in the field.

'There are some good people at The Lab but bad things are happening there right now. In the country at large. Don't bother going to the police. They are in on this too.'

In on what? It's all so frustrating. Too many questions, not enough answers. The line has gone quiet. 'Jim?' she asks. 'You still there?'

She can hear his breathing, fast and shallow.

'There's definitely someone in the house,' he whispers.

'Where are you? I'll walk round.' She senses time is running out.

'Don't – for your own safety. They'll come after you too.'

Who? The people in the Range Rover? MI5? 'Jim, please, tell me what's going on here. You're scaring the crap out of me.'

A noise on the line and then it drops.

24

Silas

Silas and Strover stand still in the car park, straining to listen for another noise. The cry of pain was unmistakable but there's only silence now. As they start to head over to where the original sound came from, a man emerges from the pub. Another familiar local. He lights a cigarette but Silas knows he hasn't just come out for a smoke. It's instinct more than anything, the way the man is trying not to look at Silas.

'Got a light?' Silas says, nodding at Strover as he walks across to the man. Silas has quit but he keeps a packet on him for emergencies.

The man offers him his lighter and they stand together, smoking. Strover heads back to the car. She knows Silas's game.

Silas rides the silence, waiting for the man to speak. When he does, his Wiltshire accent is broad and measured.

'It was parked up over there,' he says, gesturing towards Strover. 'Where your lady assistant is.'

'The Range Rover?'

The man nods. A moment later, a young woman runs down the fire escape staircase at the back of the pub and crosses the street.

'She's in a hurry,' the man says. Silas watches her disappear. Strover has clocked the woman too. 'So was the Range Rover. Left here ten minutes ago in a rush. I was having a ciggie. Nearly ran me over! But I took its number plate. Just in case, like.'

The man passes him a crumpled bar receipt with the number scribbled on the back of it.

'Thanks,' Silas says, pleased to see it begins with RO. 'Any idea where it went?'

'Straight up the hill, towards Marlborough way,' the man says.

Another pause. The conversation could go in one of two directions. More useful titbits or time-wasting.

'Something else,' the man says. 'That body being found in the crop circle.'

Silas nods. The man wants information in return, some gossip to take back to the bar, but Silas is in no mood to trade. News of the second body found tonight would send shock waves through the pub but its discovery has yet to be announced. No doubt leaks will soon start to appear on social media. It's become a problem in the force. He walks over to Strover, leaving the man to slink back into the pub.

'The Range Rover was here, left ten minutes ago,' he says, passing Strover the receipt as he opens up their car.

'I think we should take a look across the street,' Strover says, hesitating at the passenger door.

Silas pauses. It's a distraction from their investigation but he knows she's right.

25

Bella

Bella runs through the pub's archway and out onto the high street. No house is directly opposite the pub but several are set back, up the hill on the left. She crosses the road and runs down the drive of the first house. The lights are all off and she stops and listens. Silence. Doubling back, she runs down the drive of the second house, trying to get more air into her tight lungs. Should she ring the police, despite Jim telling her not to? They can't be in on it too. He's just being paranoid, ridiculous.

Someone meant us to meet.

Bella slows as she approaches what she assumes is Jim's house. Ahead of her is a small red-brick cottage, surrounded by a well-tended garden, and the front door is ajar. A cat greets her with a plaintive miaow as she pushes it open. She should definitely call the police. But a voice inside her – her dad's? – tells her to take a look around first and then make the call.

She gasps as she walks into the small kitchen. The table has been upended, the cupboard doors opened. She peers into the empty sitting room – a belly-flopped TV on the floor, a sofa on its side – and walks upstairs, barely able to breathe. Would MI5 really do all this? Are they still here? She looks around her. A closed door on the left with a light showing under it, bathroom straight ahead and a small bedroom to the right.

'Jim?' she calls out, in case she's got this all wrong. 'Jim? You still here?'

She pushes open the door on the left. It's like a bomb has gone off. The mattress is on the floor, ripped down the middle; the drawers of a wardrobe are open, clothes hanging out like entrails. No sign of Jim. She walks around, careful not to disturb anything. A desk in the corner has been ransacked. The only things untouched seem to be an upright piano in one corner and a glass cabinet in the other. It's full of vegetation and what looks like a green lizard of some kind, staring at her. She didn't have Jim down as a pianist, or a reptile person.

She should go back to the pub but instead she returns to the landing and peers into the bathroom, where a cupboard beside the mirror has been wrenched off the wall, its contents littered across the floor.

'Bella?'

She spins round. Jim is standing behind her, his face covered in blood, shirt ripped.

'Jesus, Jim! What happened? Are you OK?'

The downstairs doorbell rings. They both freeze. Jim puts a finger to his bloodied lips, signalling for her to be quiet. The bell rings again.

'We should answer it,' Bella whispers, still shocked by the sight of Jim. 'Your front door's open anyway.'

'Hello?' a male voice calls from downstairs. 'Anybody home? This is the police. DI Silas Hart, Wiltshire CID.'

'Do not tell them anything,' Jim whispers, eyes dancing. 'I'm fine, it's a superficial injury.'

'OK,' Bella says.

The cut on his forehead looks far from superficial. And why can't she tell the police what's happened? Jim would get on well with Erin. She had a similar distrust of authority, regaling Bella with stories of being chased through the back streets of Dublin by the Gardaí.

'Please,' Jim says, still whispering. 'Leave this one to me.'

26

Silas

Silas's first thought, as the bloodstained man appears at the top of the stairs, is that they were right to investigate. The tall young woman at his side seems unharmed, but the house has been turned upside down.

Two minutes later, after Strover has accompanied Bella, the woman, back across to her room at the Slaughtered Lamb, Silas sits down in the kitchen to talk to the man: Jim Matthews, aged twenty-five. Silas hasn't called the case in to the Control Room yet. The timing of the incident, so soon after the sighting of the Range Rover in the village, is bothering him.

'Are you here because of Bella?' Jim asks.

'Should I be?' Silas says, surprised by the question. Jim's cleaned up the cut on his forehead, revealing what looks like scarring from a previous injury.

'Because she's a journalist?' Jim adds, as if he should know. Silas doesn't like his tone.

'And?' Silas asks. It's sounding less like a domestic by the minute.

'You do know where I work?' Jim says, seemingly put out that Silas doesn't.

'It seems as if I should,' Silas says, checking his notepad. 'All I have is your name and age.'

'How come you're here then?' Jim asks.

'We heard a noise,' Silas says, looking around at the small kitchen, the pots and pans on the floor, the open cupboards. 'Someone crying out in pain.'

Something's not right. It reminds him of the ransacked scene they found in Noah's house, but it feels different, almost as if it's a film set.

'And you just happened to be passing?' Jim asks.

'That's right,' Silas says.

Jim smirks. Why's he so bloody suspicious of him? Most people around here complain there are not enough police in the countryside.

'Tell us about this woman, then,' Silas continues, checking his notebook again. 'Bella. The journalist.'

'She approached me in the pub tonight. Everything I told her was OSINT.'

Silas looks up, intrigued by Jim's choice of words. 'You better tell me where you *do* work,' he says. 'OSINT' is the language of spooks and means open source intelligence – unclassified information that can be found in the public domain.

Jim looks around the room, seeming to enjoy the suspense before he turns back to Silas. 'Porton Down – I'm a government scientist. Chemical analyst by day, amateur mathematician by night.'

Silas's expression remains unchanged, but alarm bells are already ringing. Not because of any potential breaches of national security but because of Ward's earlier words, warning him to tread carefully with Porton Down. If Jim's in some way involved with the crop circles, Silas's life has just got a whole lot more complicated. A nightmare of paperwork and Official Secrets Act protocols, Ward breathing down his neck at every turn, liaising with the funny brigade at MI5. His eye is caught by a book on the floor. Oh Christ.

'We just had a drink,' Jim continues. 'Then she told me she was a journalist. I was pretty shocked – we're warned regularly to watch out for approaches from the media.'

'But you still invited her back here?' Silas asks, mustering a tone of surprise. He glances again at the book.

Jim shakes his head. 'She was staying at the pub. I called her when...'

Jim pauses, flustered for the first time in their chat.

'When what?' Silas asks.

'When I heard someone in the house. She kindly came over to help. I asked her not to but she did and found me like this. She's innocent. I've told her nothing of consequence about the work I do at Porton.'

Silas sits back. What's really going on here?

'We happened to be in the village because we were looking for a car – a Range Rover,' he says. 'Don't suppose you saw it this evening?'

'I didn't, but Bella did. In the pub car park. We'd gone our separate ways – after she'd told me she was a journalist – and then she returned, said there was a Range Rover outside. She wanted to warn me.'

'Warn you?'

Jim nods. 'I'd told her that it had nearly driven me off the road earlier, on my way home from work. It'd been on my tail for a while.'

'And you think the same people were responsible for what happened here?' Silas asks, unable to disguise his scepticism.

'What I do at Porton is sensitive work, inspector. I heard someone in my house tonight. Just as I was going to sleep. It sounded like they were searching for something. When I went downstairs to investigate, they must have hit me on the head. I'll report the incident tomorrow to my team leader, when I get into work. Nothing's been stolen. I never bring back classified documents from the office anyway. Some do, not me. I'll also declare that I was approached by a journalist. No doubt security and the press office will be informed.'

Silas has no interest in the journalist, hasn't got time. He needs to focus on a possible link with the Range Rover and the crop circle victims. There are too many echoes of what happened to Noah.

'One last question.' Silas picks up the book from the floor. The glossy cover is an aerial shot of a circular geometric pattern in a field of golden wheat. This time it's his turn for a dramatic pause. 'What's your interest in crop circles?'

Silas scrutinises Jim for a telltale reaction as he hands him the book, but he seems unfazed by the question. 'I enjoy the mathematical ones,' Jim says, flicking through the glossy pages.

'Like Euler's Identity?' Silas offers.

'You know it?' Jim is more delighted than surprised, a sudden childish smile lighting up his face.

'The crop circle that depicted it had an exquisite beauty,'

Silas says, bluffing. He's out of his depth and should stop. Right now. 'So I'm told.'

'Like the equation itself,' Jim enthuses. 'In technical terms, it shows a deep relationship between the trigonometric and complex exponential function.'

Silas nods but he hasn't a clue what Jim's on about. 'I'll take your word for it,' he says, pulling out a police drone photo of the first crop circle killing. Happy to be on more familiar territory, he hands the photo over to Jim. 'And what do you think that is?' he asks, again scrutinising Jim for a reaction. 'Not quite so beautiful.'

It's a while before he answers. 'Why are you showing me this?' he asks, looking up at Silas.

'You're a scientist. I thought you might know what it means.'

Again, Jim takes his time before he replies. 'Isn't it obvious?'

Silas looks up. Jim's tone is more baffled than patronising, as if he genuinely can't understand why Silas doesn't know what the crop circle means. Silas takes the photo and pretends to give it a second look, but he's desperate to know more. 'Not to me, it isn't.'

'The binary wheel is an encoded chemical formula,' Jim says, hesitating. 'And the hexagons – that's its molecular structure.'

'Structure of what, exactly?' Silas prompts.

'3-Quinuclidinyl benzilate, otherwise known as BZ or Agent 15, an incapacitating chemical warfare agent stored in the vaults of Porton Down.'

27

Bella

'Thanks for your time,' DI Strover says, standing at the door of Bella's room in the Slaughtered Lamb.

'No worries,' Bella says. The detective has been gentle with her, once she'd established that Bella wasn't responsible for Jim's injuries.

They trade farewell smiles, but Strover lingers at the door.

'Sure you're OK?' she asks.

'I've got a friend,' Bella says, blinking back hot tears, 'Erin. Her dad was beaten up by her mum.'

'It does happen,' Strover says.

'She was very young at the time,' Bella says, more together now.

'Your friend?'

Bella nods, turning away. 'Six, I think.'

'You've got my number if anything else comes up,' Strover says. 'Look after yourself.'

Bella smiles but as soon as the detective has gone, she sits down on the edge of the bed and the tears flow freely. It's been a crazy few hours, culminating in the sight of Jim appearing on the landing of his house, covered in blood. She hopes he's OK. As Strover led her away, Jim gave her a reassuring smile from the kitchen sink. Hart was less friendly, his eyes full of accusation as she left the house. Did he really suspect her of attacking Jim?

Bella gets up from the bed and looks down on the almost deserted car park. Strover and Hart appear to her left, walking across to their unmarked vehicle in the corner. Strover glances up at her room and a sudden chill runs through Bella. She turns away from the window. The only other time she's been interviewed by the police was pre-Oxford, another life.

She wants to go around to Jim's house and see how he is, compare notes about their interviews, reassure him that she didn't say too much to the detective. After she'd realised that Bella was an innocent party, Strover seemed more interested in the Range Rover, kept asking about its appearance in the pub car park, why she'd warned Jim. Bella didn't reveal that she knew he worked at Porton Down, or that he thought MI5 was following him. She didn't want to get him in trouble – or do anything that might jeopardise a big story.

She pulls out her phone and calls Jim but it goes straight to voicemail.

'Hi, it's me,' she says. 'Hope you're alright and it went OK with the police. Call me when you get this and I can pop around again if you like. Just seen the detectives drive away.'

She daren't say any more in case her phone's being bugged. Jim's paranoia is contagious. If only he'd elaborate, at least give her a clue about Porton Down and its possible connection with the crop circles.

She doesn't want to worry her mum so instead she tries Helen in Australia. This could be her big break in journalism and she'd like to share it with her older sister. It's now morning in Sydney and Helen should be at home, but the call also goes to voicemail.

'Hey, big sis, it's me.' Bella pauses, fighting back a second wave of tears. She's always called her 'big sis', even though Bella's been taller than Helen for as long as she can remember. 'Ring me when you get this. Not sure if you've seen the news from here, but a body's been found in a crop circle in Wiltshire. It's a mystery but your intrepid reporter is on the case – I think I know who's behind it. Dad would be proud, I hope. It's my chance, Helen. A potential splash. I just need to persuade my main source to tell me more.'

28

Silas

'Well, that wasn't what I was expecting,' Silas says, as he drives out of the village with Strover. 'Sounds like Jim was followed by the Range Rover – on his way home from his day job at Porton Down.'

'Porton Down?' Strover asks, surprised. 'Bella never mentioned that.'

'It's all we need,' Silas says. 'A chemical analyst, apparently,' he continues. 'And an amateur mathematician.'

'Interesting combination,' Strover says, pausing. 'I think she quite fancies him actually. Can't say I blame her.'

'Really?' Silas gives her a sideways glance, thinking back to the oddball he's just interviewed.

'Like a young Jeff Goldblum,' Strover says, looking out of the passenger window.

Silas shakes his head. No accounting for taste.

'Either that or she was just protecting her source,' Strover continues.

'What if he's a croppie, worked with Noah in some way?' Silas asks.

'You think he's a suspect?' Strover looks at Silas with surprise. 'I thought we were interviewing them about possible domestic abuse.'

'We were,' Silas says, pausing. 'But something wasn't right about him.'

He thinks again about the book on Jim's floor.

'He had an unhealthy interest in mathematical crop circles – formulas. Which is too much of a coincidence, don't you think, given he was followed by the same Range Rover that visited Noah's house? And he's certain the first circle contains the formula for an incapacitating chemical warfare agent called BZ. Also known as Agent 15.'

'My Cambridge mathematician would have picked that up,' Strover says, but she makes a note of BZ and Agent 15 and underlines them both in her pad.

'I don't doubt it,' Silas says, backing off. 'Jim seemed so bloody paranoid, worried that he might lose his job because he'd met a journalist in the pub.'

'Is that such a crime?' Strover asks.

'To be fair, it didn't exactly do David Kelly much good.'

'David who?' Strover asks.

'A scientist who used to work at Porton Down,' he says. 'Before your time. He made the mistake of speaking to a journalist about the sexed-up dossier on Iraq's weapons of mass destruction. Two months later, Kelly was found dead on a hillside in Oxfordshire.'

'You going to call this incident in to Control then?' Strover asks.

'I should flag it up with the MOD Police at Porton,' he says, 'in case Jim doesn't report it himself, but I'm going to leave it for a while. It might buy us some time with Ward.

I don't want him all over this, telling us what we can and can't investigate.'

It's a risk and Silas can tell from the silence that Strover's not happy with his decision. She's still early in her career, playing by the rules. They drive on through the darkness towards Swindon, home for both of them. Strover lives in a terraced house close to the train station. He and Mel have moved out to Blunsdon, a quiet village that looks north to the Cotswolds and offers the promise of escape from Swindon.

'So what do you reckon the connection is between our apparently "handsome" government scientist and the Range Rover?' Silas asks, still baffled by Jim's attractiveness to women. He enjoys his time in the car with Strover, the chance to share his policing experience, the way she keeps him young, in touch. It's why they're a good team, the envy of the station.

Silas's phone rings before Strover can answer. It's the Control Room with news he doesn't want to hear at any time, let alone at this hour of the night.

'Sir, we're getting reports of a third body found in a crop circle.'

29

Bella

Bella stands outside Jim's house and knocks on the door, moving back to look up at the landing window. The curtain's closed but there's a light on inside. She can also hear a piano being played. Jim hasn't returned her calls and she wants to talk to him about the police interview, how much he told them. Did they turn up tonight because he'd met a journalist in the pub and they thought he was leaking a story? If that was the case, surely Strover would have asked Bella more questions about her evening with Jim, his work. Her work.

She knocks again but there's no response. Jim's piano is upstairs. She saw it when she came to the house earlier. He won't hear her if it's him playing. She stops to listen. A classical piece, formal, ordered. Glancing around her, she walks down the side of the house to the back garden. The light is on in Jim's bedroom, a net curtain hanging across the closed window. She walks further away from the house to get a better view, picking her way carefully through a tidy rockery. Everything is in its place, neat patterns of brick paving and circular flower beds.

If she stands on tiptoe, she can see Jim's silhouette at the piano. He's playing with controlled passion, back ramrod straight, hands moving fast. She pulls out her mobile and tries him again. Voicemail. He must have turned off his phone. Should she try to throw a stone up at the window? He was jumpy earlier, nervous. It might alarm him too much. Instead, she watches him play, listening to the beautiful music drift out into the summer night. Jim seems so rational, systematic – it must be the scientist in him – but there's an artist's passion about him too. He's also calm, reassuring company, until something slips and he's on edge like a wild animal: wide-eyed, alert.

After five minutes, Bella begins to feel cold. She should head back to her room, try Helen again. A deer barks in the woods behind her, a deep, guttural sound. There are no cars on the road and the village is still. Too still.

Jim stops playing and the silence is almost deafening. It's hard to see him clearly through the net curtain, but she can make out his profile, standing now. What's he doing? She blinks, barely believing what she's just seen. Jim seemed to walk forward and hit his head against the bedroom wall, stumbling backwards from the impact. Did that really just happen? She looks up again, watching as he walks towards the wall for a second time and smacks his head into it, the blow so hard he falls over.

Bella turns away, sickened by the sight, but she can't help but look back up at the window again. Jim's up on his feet, approaching the wall for a third time.

'Jim!' she calls out into the night sky. 'Please, it's me, Bella.' But Jim doesn't hear her. Somewhere a sash window slides open. She braces herself for abuse from a neighbour, but there's only silence. Jim approaches the wall for a fourth time, moving like a malfunctioning automaton. She can't bear to watch any more and slumps down onto an ornate wrought-iron bench, holding her head in her hands.

Erin used to bang her head like that at college, when she was really out of it. Over and over until the porters came. Bella begged her to stop but she would never listen, too high on whatever she'd taken that day. She'll text Erin when she gets back to her room, try to cheer her up – assuming she is at least receiving her texts and voicemail messages. Remind her friend of the time they summoned a porter to Erin's room on the pretext that she had an intruder. As soon as he arrived, Erin slipped out the door and they managed to shut him inside.

It was the first bad thing that Bella was involved in at college. They stood outside for ten minutes, giggling as they listened to him knocking on the door, politely at first and then more urgently. He was a nasty piece of work, cruel to all the students. Bella let him out in the end – Erin would have thrown away the key. Silly high jinks.

Bella glances up at the window to see Jim walking past as if nothing has happened. Did she imagine the whole thing? She looks around, soothed by the fragrance of philadelphus behind the bench. Their garden in Mombasa was full of scented flowers: bougainvillea, frangipani and orange jessamine. Her mum used to get her and Helen to draw them and write their names underneath. She misses Helen so much. Erin too. Why can't she see her friend?

She turns to walk back to the pub. Maybe Jim will answer his phone now that he's finished playing the piano.

And stopped banging his head against the wall. What was that about? In the distance, a roosting bird flies up into the night sky, angry and incongruous, its dark body silhouetted in the moonlight.

30

Silas

Thirty minutes after the call from the Control Room, Silas walks up a field outside Stanton St Bernard with Strover, holding an old station flashlight that he keeps in the car. It belonged to his dad, who was also in the force. Ahead of him is a single uniform. It might be the fading light from the old torch, or the cold that's descending on the summer night, but the young officer looks like he's seen a ghost.

'You alright?' Silas asks, glancing at the flattened wheat behind him.

'Not really, sir,' the officer says.

Silas flashes the torch across the field, fixing the beam on a figure slumped on the ground. This body hasn't been carefully placed, like the others. The person has clearly been involved in a struggle of some sort, legs splayed at an awkward angle, arms flung wide.

'Who found it?' Silas asks, walking towards the body.

'Dog walker,' the officer says. 'My colleague took her back to the station.'

And left him here on his own, the poor sod. He looks young enough to be at school.

Strover hangs back, chatting to the officer as Silas approaches the body. Above him, a spill of stars illuminates the big Wiltshire sky.

'Sir,' Strover calls out, but Silas ignores her, tries to ignore a rising fear in his stomach.

The body is lying on its back, head turned away from him. He keeps the flashlight trained on the figure as he moves around and squats down to see the face. The victim is male and he's wearing dark trousers and a loose T-shirt.

'Sir,' Strover calls out with more urgency.

Strover should know better than to disturb him when he's inspecting a corpse. Silas shines the light on the body's ashen face and starts back in shock. Part of one cheek has been chewed away, revealing an upper row of teeth. A deer maybe. Or a fox. He's used to dead, violated bodies, shouldn't be feeling the coldness that's crept up his spine like a fast-rising tide, but something is different. What is it, apart from the flesh wound? He's much older than the other two victims, a similar age to Silas, and the dead eyes are wide open, staring ahead with an empty finality that he's seen in the deceased too many times before. And then he realises. There's a pulse at the man's neck, the grey skin twitching faintly.

'Sir,' Strover calls out again. 'The victim's still alive.'

Silas stands up and backs away, the light of his torch still on the man's eyes. Jesus Christ. He looks around at the flattened wheat, which has been ruffed up in several places near the body. What the hell's happened here? The other scenes were disturbing but they were places of calm too. Of final rest. He forces himself to walk back over and kneel down beside the body again. Just to be sure, he puts his hand on the victim's neck, looking at the open wound, the congealed blood. The pulse is slow but steady.

'Can you hear me?' he asks, leaning down to speak in the man's ear. 'My name's DI Silas Hart. Swindon CID.'

The man stares ahead, his face expressionless. Is he in a coma? On spice, perhaps? The synthetic marijuana that's become so prevalent in the UK can induce a similarly catatonic state.

'Blink if you can hear me,' Silas says, watching for a reaction. Nothing.

'An ambulance is on its way,' Strover says, now at his side.

'I've never seen anything like this,' Silas says, but Strover has turned away. She too is struggling to cope.

Silas reaches out to touch the man's forehead with the back of his fingers. Cold as ice. The wound is shocking but it's the man's eyes that are troubling him. A wide-open look of fright, frozen in time. What did he see? Can this man really be alive? The neck is still pulsing, a distant echo of life. He must be close to dying of hypothermia. Silas takes off his jacket and lays it across the man's torso, checking his legs and arms in case there is any blood or injury. And then he notices a gash on his left arm. It's not on the underside, but on the back of the forearm, a deep angry slash of a wound. Fresh too but maybe not as recent as the cheek wound.

Silas doesn't believe in a God but as he rises unsteadily to his feet, he offers up a quiet prayer for this man's soul, if it hasn't left the body already. Noah had raised two fingers.

He must have meant two more crop circles. At least there are no more. It's the only consolation as he joins Strover and talks to the officer, who still looks like he's going to pass out. In the distance, the sound of a solitary ambulance driving up through the moonlit vale.

31

Jim

Jim has tidied up the mess in his house, practised the piano, made himself a hot chocolate and fed his pets, but he still can't sleep. Not after what's happened tonight. First the discovery of the crop circle and then the visit from DI Hart. His head's sore too after the intruder. The detective seemed to have no knowledge of the Range Rover that had followed him back from work, but it was too much of a coincidence: him just turning up at the house with his female colleague like that.

Maybe they'd been following Bella, knew that she was going to approach him in the pub. A part of him thinks the encounter might have been a set-up, a classic spy sting, to see if he would tell all about Porton. In which case, Bella is also in on it and can't be trusted either. She's certainly attractive enough to be a honeypot, as they were once quaintly known. The Russians call them swallows – intelligent women who use their beauty to encourage betrayal.

She's rung a few times but he's let the calls go to voicemail while he makes up his mind about her. It was good to meet someone again but he doesn't trust himself, the speed of their blossoming friendship. The last time he went out with anyone was at Warwick, in his final year, and she broke his heart.

He turns on his bedside light, reaches across for his laptop and searches again for news of the crop circle. An article on what the coded symbols might mean catches his eye and he pulls up an aerial photo, studying the distinctive hexagonal patterns and the adjoining binary wheel. Various theories have been put forward by armchair pundits around the world but they've all missed what they depict: BZ.

Jim hasn't tried to decipher the circular pattern properly. It's been divided into cake slices of eight units – each one depicted by a small patch of flattened or standing wheat. Which suggests eight-bit binary sequences that he's sure will convert into ASCII-generated text – and the chemical formula for BZ.

As for the adjacent collection of hexagons and other shapes, they clearly represent its molecular structure. Two in particular – big flattened areas of wheat with double-lined borders along several sides – look suspiciously like benzene rings. It could be something else, of course, but he's got the same feeling of confidence that saw him through his finals at uni. Even Dad was chuffed with his first, one of the highest the university's ever awarded. Porton Down was impressed too – offered him a job on the spot.

He looks again at the adjoining circle and its binary sequence. Someone out there is sending him a message, encouraging him in his mission to expose the truth about Porton Down. He thinks of all the classified information he's managed to access since being transferred back from

Harwell at the end of last year. BZ, also known as 'buzz' as well as Agent 15, was originally developed as an ulcer therapy but it proved unsuitable. The smallest dose could induce a severe psychological storm, total mental mayhem, which appealed to the American military, who were looking for incapacitating agents, or 'humane' weapons, in the 1960s that could disorientate rather than kill the enemy. After being used against the Vietcong in the Vietnam war, BZ was allegedly stockpiled by Saddam Hussein in Iraq and, according to Syrian rebels, deployed by the Assad government against them in the siege of Homs. More controversially, the Russians claimed that BZ, rather than novichok, was used to attack the Skripals in Salisbury.

The powerful hallucinogen or 'zombie gas', as it's been dubbed, is still stored at Porton Down, officially for defensive rather than offensive purposes, but Jim knows better. He tilts the screen of his laptop, admiring the strange symmetry of the hexagons, their relationship to each other. Downstairs his dad's grandfather clock strikes one. It's late. He needs to sleep. But he can't resist a quick search on Twitter. No one appears to have made the connection with BZ yet. It was a mistake to have told DI Hart earlier. Too much of a risk. He hasn't even told Bella.

A new tweet appears and his eyes start to blink. Rumours are circulating of a second crop circle that's been discovered by police – with another body in it. The police are shortly to make an announcement. Jim searches in vain for photos. It's too late, too dark, for anyone to have taken pictures.

He doesn't need to see it though. He can guess exactly what the second pattern will depict. Lysergic acid diethylamide. Otherwise known as LSD. Acid. He's sure of it. Another message of encouragement to him. Another substance interwoven with the grim history of human experiments at Porton Down. Most famously, a troop of Royal Marine commandos were given LSD in December 1964 as part of an experiment to see if it too could be deployed as an incapacitating agent. The marines were sent out on exercise on a nearby firing range – 'happy valley', as it's since become known – but within the hour, they had dissolved into a bunch of giggling fools.

What no one seems to realise is that such unethical experiments are not confined to Porton Down's controversial past. After his three years at Harwell, Jim knows the truth, and now someone else on the inside does too. Someone who seems to be encouraging him to act.

32

Silas

It's 1 a.m. by the time Silas slides the key into his front door. The house smells of flowers – Mel has a big wedding tomorrow at Rockley, a hamlet outside Marlborough, and the kitchen floor is covered with buckets of freshly cut tuberoses. *Polianthes tuberosa*. She's been teaching him their names, Latin ones included. He pours himself a whisky and inhales the sweet scent all around him. It couldn't be a better antidote to the mutilated body in the field he saw tonight, the dank miasma of death. Except that it wasn't.

He's about to creep upstairs when he notices the light on in the sitting room. Mel usually goes to bed early if she has a big day coming up but she's in the armchair beside the fireplace, an empty whisky glass in her hand, eyes closed.

'I thought you'd be in bed,' Silas says quietly, leaning down to kiss her on the lips.

She opens her eyes and smiles, noticing the glass on her lap. She puts it on the table, her actions slow and sleep-heavy.

'I wanted to wait up for you,' she says.

'But you've got a fancy wedding in the morning,' Silas says, sitting down in the armchair opposite.

'You only get back this late when something really shit's happened at work,' she says. 'And when it's that shit, you need to talk about it.' She pauses. 'That was our problem in the past.'

Silas closes his own eyes – he's lucky to have Mel – and takes a sip of his whisky.

'Want another one?' he asks. She nods.

A minute later, after fixing her a whisky and topping up his own, he tells her about the body in the field, the half-eaten cheek, the look on the faces of the paramedics who took the man away. He's never told her the details of cases before and he's not sure it's the right thing to do now, but it feels good to talk.

'They were so shocked he was still alive,' Silas says. 'Everyone was. The low body temperature made no sense, there was almost no blood pressure, a barely beating pulse.'

'He's still alive now?' Mel asks.

Silas nods. 'The hospital rang me on the way home. No improvement but at least he's breathing. God knows how. We just need to find out who he is.'

Noah is still at the Great Western Hospital too, yet to regain full consciousness. He'll visit him in the morning. Silas pauses, thinking again about the man in the field. 'It was like he was a zombie, or something.'

'A zombie? You've been watching too many films with Conor.'

Conor is out tonight with Emma. It's the first time he's been in a relationship for years. When he's at home and not with her, he's watching horror movies. Religiously. Silas has tried to sit through a few of them with him, all part of what the counsellor calls 'reconnecting', but it's not the best way to unwind after work. He prefers comedy horror. *Shaun of the Dead*, that sort of thing.

They say nothing for a while, but Silas doesn't mind. The quiet is soothing, far removed from the accusatory silences that used to hang above them like storm clouds.

'Maybe that's the intention,' Silas says after a while. 'The final part of the message. First a lobotomy, then a straitjacket. And now a…'

He can't bring himself to say the word again. The Hollywood portrayal of zombies, even the darkest ones, doesn't bear any resemblance to what he saw – and felt – tonight. The surprise of seeing the man's teeth, not where they should be, glinting in the torchlight high up in his torn cheek. The pale stillness of his skin. The empty, unlit eyes. This man's soul was troubled, had wrestled with something awful.

'You alright?' Mel asks. Silas realises the whisky glass in his hand is trembling.

'I'm OK,' he says.

Ten minutes later, Mel holds him tight in bed as he tries in vain to switch off and sleep.

'It's good about Conor, isn't it,' Mel says. 'Where he's at now.'

'He's done well,' Silas says, feeling guilty that he hasn't given their son's turnaround much thought recently.

'We must be prepared for a relapse, not panic,' she says, almost asleep. 'If it happens, we deal with it. Together.'

'He's good now,' Silas says, but his mind is wandering. Jim's confident pronouncement that the first circle was a chemical warfare agent has rattled him, particularly after Ward's visit to the crime site and his warning about Porton Down. Strover's checked the symptoms of BZ too, which include sedation leading to stupor – not so different to how it must feel after a lobotomy, the procedure that the first victim was subjected to. He should check in with the MOD Police at Porton Down tomorrow, ask them about Jim, but that will mean telling the boss what he's up to, risking the entire investigation. Ward's closed him down before.

He thinks again about the latest victim, his age. So much older than the other two. Who is he? There was no fingerprint match. DNA results will come back tomorrow. He had no ID on him either. The only possession they found was a single old key in his pocket. CSI have sealed off the site and will continue to comb the surrounding area at daylight. They need to establish how he got there. The location is even more remote than the one at Chute Causeway.

'Silas, you need to sleep,' Mel whispers. 'I can hear your brain whirring.'

He closes his eyes but a nagging thought won't go away. Whoever placed the first two bodies in the crop circles knew they were dead, knew they couldn't talk or reveal anything about how they got there. The third man was left alive in the field. He might die, but he might also survive – and disclose how he got there. That's a big risk. Too big a risk for someone else to take. Which makes Silas think that the man they found tonight was meant to be found alive. And the key was left on purpose too.

Silas listens for Mel's breathing. It's regular and shallow, the sleep of someone at peace. She's become a different person since retraining as a florist, found her thing. And become much nicer to Silas. He lifts up the duvet and slides out of bed, careful not to wake her as he tiptoes from the bedroom.

Downstairs, he pours himself another whisky. There's no way he'll sleep tonight without more alcohol in his blood. Unless he can explain why the third body was left alive on the hillside. It's driving him mad. He walks through into the sitting room in his striped pyjamas – a recent present from Mel – and looks out onto the lawn through the French windows. Mel teases him but the edges definitely need trimming. A tiny bit of order in a chaotic world. His dad's lawn always had tidy edges.

Silas turns away, sipping from his drink. Why would someone leave a man like that, with the risk of him surviving? Everything about him is different from the first two victims. They weren't just dead – they'd been dead for a long time. Possibly at the hands of someone else, if Strover's right.

And then he understands.

'I think it's him,' Silas says, slipping back into bed. He shouldn't wake Mel but he can't help himself.

'What?' Mel asks, surfacing from sleep.

'He placed the first two bodies as messages. And then he left himself.'

33

Jim

Jim turns up the music as he drives out of the village, first over the railway and then the canal, wrapped in a light veil of mist beneath him. It's dawn and he's hardly slept but he's worried about staying at home. The crop circles are dominating the news following the discovery of not one, but two more bodies in the middle of elaborate patterns.

Somehow, TV news stations have managed to get early morning photos too. The second circle, as he suspected, looks suspiciously like the molecular structure of LSD. And the third circle bears more than a passing resemblance to VX, one of the world's most deadly nerve agents, a supply of which is also kept at Porton Down. Most recently, VX was used to murder Kim Jong-nam, the brother of North Korea's ruler, in a very public assassination at Kuala Lumpur international airport.

Jim tries to relax as he heads along the Fair Mile, an old Roman road that's a favourite part of his morning commute. When he did snatch some sleep, it was light and his dreams were of Bella. For some reason known only to his deep, unconscious mind, she was naked and riding a giraffe. He shakes his head, smiling at the memory. Eat your heart out, Freud. He's been dreaming a lot of weird stuff recently. And then he thinks of Bella in the pub last night, blue-grey eyes beneath a stern fringe. And glasses almost as big as his own. Bookish but less awkward than him. One minute studious and detached, the next all blushes and warmth. But he senses that good humour isn't her default demeanour. It has to be earned. Maybe he's flattering himself.

He'll phone later, apologise for not returning her calls last night. He's decided that their encounter in the pub wasn't a sting. There are, however, some unanswered questions. Why did her arrival in the village coincide with the discovery of the crop circles? And why did she choose to sit at a corner table in the pub – his table? But his normally cautious self has overruled them and he's listened instead to his heart.

He turns up the music on the car's sound system, conducting with one hand. Bach's 'Goldberg Variations'. And Glenn Gould, his idol, is at the keyboard. Effortless, lucid, pristine playing. Listen closely and his humming is audible too. One day Jim hopes to play Bach's masterpiece as well as Gould. He likes the work's mathematical symmetry, the way the second voice in each canon kicks in a note higher until the eighth canon, when it drops to an octave lower, creating circular patterns – geometric shapes in sound.

The fourth aria is interrupted by his mobile ringtone. No caller ID. He ignores it as he turns right at the Shears Inn and cuts down towards the A338 to Salisbury. The anonymous caller rings again. This time Jim answers it.

'Who's this, please?' he says, waiting to join the main road.

'The car behind you.'

Jim shoots a glance in the rearview mirror. On cue, a black Range Rover appears around the corner. It must have followed him at a distance on the Fair Mile, hidden itself in the dips, or he wasn't checking.

Jim hangs up and accelerates away at the junction, heading south towards Tidworth. His mobile phone rings again. Damn it, should he take the call, speak to these people? It's better than being driven off the road. They must be MOD Police and want to talk to him, that's all. He shouldn't have sped away yesterday. He'd nearly crashed. Besides, they can't pin anything on him.

'Can I help you?' Jim says, trying in vain to sound calm.

'How did you meet Bella?' the man on the phone asks. No introduction, no manners. Just a straight question. Dad always taught him the importance of good manners. Has he heard the voice somewhere before? Jim glances again at the mirror. The Range Rover is three cars back.

'I thought you might have sent her,' Jim says, doubts resurfacing. He's still certain he and Bella didn't meet in the pub by chance.

'Did she approach you?'

Jim thinks back to how they met. Maybe she did. She was sitting at his table, waiting. He checks the mirror again. No need to land her in trouble. He likes Bella, quite fancies her, if he's honest. He just wishes he could know for certain that he can trust her. At least she came clean, told him she was a journalist. She wouldn't have done that if it had been a sting.

'Someone made a mess of my house,' Jim says, changing the subject.

'What sort of mess?'

'Come on, guys, you're wasting my time here,' Jim says, tapping his fingers on the steering wheel. He doesn't need to put up with this. 'One of you sat in the car park while the other trashed my place, looking for something. I never bring classified work home with me. What you did was pointless. And I told her nothing. I've explained all this to your colleague, DI Hart, if that's his real name.'

'Who's DI Hart?' the man asks. For the first time there's a tone of surprise in his voice. Maybe concern too.

'You tell me,' Jim says. 'Swindon CID, apparently.'

'What did he want?'

'Same as you. This conversation is over.' Jim leans forward, his finger hovering to end the call. 'I'm going to work and will report what happened last night to my team leader. End of.'

'If you can pull in to the lay-by up ahead, we just want a chat,' the man says, making a pathetic stab at friendliness.

Jim hangs up and puts his foot down. The road ahead is clear. At the A303 roundabout, he turns left, instead of continuing down on the Salisbury road, and takes the first right towards Palestine and the Wallops. Hampshire does a good line in village names.

The road narrows but he knows it well and maintains his speed, checking the mirror every few seconds. Has he lost them? The distinct profile of the Range Rover appears in the rearview mirror again. Jim gives the steering wheel another smack. And then he sees a tractor in a lane to his left, trundling down to join the road on the corner.

Jim often comes this way when the main Salisbury road is busy. Slowing, he lets the Range Rover draw close. On cue, his phone starts to ring again. He glances over at the tractor, now within twenty yards of the road. A quick calculation and he flashes his lights, ignoring the phone. The tractor driver assumes Jim is letting him turn and pulls out in front of him, but at the last moment Jim nips through the gap between the tractor and the hedge.

A loud blast on the tractor's horn. Jim speeds off on the far side, glancing in the mirror to see the Range Rover slide into the nose of the tractor. It too had tried to accelerate through the gap but must have failed – as Jim had hoped. He slows, still watching, a sickening feeling in his stomach. The driver jumps down from the tractor but no one else is moving. Airbags have deployed in the stationary Range Rover. Before the scene disappears around a bend, Jim can just make out the tractor driver on his phone, gesticulating frantically in the road.

34

Bella

Bella's not big on breakfast, but as it's included in the price of her stay at the Slaughtered Lamb, which she might be paying for herself, she settles down in the bar and tucks into a bowl of Bircher muesli. The corner table, where she sat last night with Jim, is empty, but she can't bring herself to sit there. Instead, she glances over at it, wondering if any of last night actually happened.

She tried calling Jim first thing this morning but his phone rang and rang. And when she went round to check on his house, it was shut up and his car had already gone. The image of him banging his head against the wall still haunts her, partly because she can't be sure it happened. It was almost like a ghastly puppet show, his silhouette barely visible through the diaphanous curtain.

'Another coffee?' the waitress asks, as Bella glances through her notes from last night.

'Thanks,' Bella says. 'Are there any newspapers?'

'Which one you after?' the waitress asks. If she's surprised that a twenty-one-year-old wants to read a newspaper, she doesn't show it.

Bella asks for a copy of the paper she works for, glancing at the other guests who are eating their breakfasts. A retired couple in the corner has the same paper on their table.

'Here we go,' the waitress says, lingering as she looks at the front-page headline. 'Nasty business, these crop circle killings.'

Bella glances at the image. It's an aerial photo of a crop circle with a police forensics tent in the middle.

'Apparently they found two more bodies overnight,' the waitress adds. 'Just near here.'

'Two more?' Bella asks, looking up.

'So they're saying on the news.'

'Are they linked to Porton Down, do you think?' Bella asks, skimming through the text.

'Who knows anything these days?' She pauses. 'Saw you having a drink with our foxy Palmer last night. He'd know if anyone does.'

Bella blushes as the waitress walks away. She hopes she and Jim will meet again, and not just because he's got a story to tell.

'You've an eye for the loo-lahs, for sure,' Erin had once joked, after Bella had kissed a crazy boy at college. A porter had caught him running around First Court naked in the snow – one of many college rituals – and he was being escorted back to his room in a blanket, singing Nick Cave's 'Fifteen Feet of Pure White Snow'. The relationship never came to anything – he dropped out of uni shortly afterwards.

Bella starts to read the paper when her mobile rings. It's Jim.

'Hello?' Bella whispers, glancing around the pub. She doesn't want the waitress to know she's talking to 'foxy Palmer'.

'I might have to disappear for a while,' Jim says. He sounds different, focused in a frightening sort of way. 'Sorry I didn't call back last night.'

'Are you OK?' Bella asks, getting up from the table. Her legs have turned to jelly but she manages to walk out of the pub into the fresh air. Jim's in his car and reception is poor.

'There's a key around the side of my house, under the flowerpot beside the staddle stone,' he says. 'It's for the back door and I want you to go in and take Rocky over to the pub. I know it's a big ask but I didn't realise I might not be returning. They always look after him when I'm on holiday. Tell them I've had to go away for work. The cat will hang out at the pub too but you can't make her.'

'Jim, who's Rocky?' Bella asks, her mind racing. Did she miss someone at the house? An elderly relative, perhaps?

'Sorry, Rocky's a young Yemen chameleon. I didn't show him to you last night. It wasn't really the time for introductions, was it?'

'OK,' she says, remembering that she had seen it briefly last night. 'Why aren't you coming back, Jim?'

'I can't explain now. They followed me again to work. The people in the Range Rover. Do not trust the police if they get in touch again, do you understand?'

'I get it.'

'I repeat, the police are not your friend.' He pauses. 'Something's happened, Bella. And I have to stay low for a while. Please listen carefully. Under the stone, in Rocky's vivarium, there's a USB stick. It's encoded but take it. Rocky won't bite, I promise. He's more into locusts. I'll let you know how to decode the USB later. Everything's on there – the truth about Porton Down, these crop circles. I must go now.'

'Jim, wait,' Bella says, glancing around the empty car park. The line falls silent but he's still there.

'I really enjoyed meeting you last night,' she says. 'Not just because I'm a journo and you work at Porton Down and all that but, you know, because I felt... we had a real connection.'

More silence. 'Me too,' Jim says, and this time he does hang up.

35

Silas

'Any word on the Range Rover's number plate?' Silas asks, sitting down next to Strover in the wrong corner of the Parade Room. A bad start to the day has just been made worse by the arrival of a contingent of French police officers on a cultural exchange. Silas loves the French, loves France. He's just not so keen on the Wiltshire Police liaison officer who told them, in the new spirit of hot-desking at Gablecross, that they could sit in CID's preferred corner.

'Looks like it's cloned,' Strover says.

'Another bloody show plate,' Silas says, sighing.

A baffling DVLA loophole that should have been closed years ago. Police are seeing an increasing number of show plates – number plates that are meant for off-road use only and can be bought without documentation or proof of address. As a result, criminals are able to buy the plates for already registered cars and use them on stolen vehicles. Once cloned, the cars appear legit – and the original owner gets all the speeding tickets.

'I checked in with my Cambridge academic this morning,' Strover says, changing the subject.

'Did you ask her about BZ?'

'She's still discussing it with the chemistry professor at Imperial,' Strover says. 'The molecular structure of BZ certainly contains two hexagons, just like the first crop circle, but the rest of the adjoining shapes are harder to decipher. On the plus side, there are similarities between all three circles. If we can crack the code for one, it should unlock the others. But the patterns are quite crude approximations, done in a hurry.'

'They can't be the easiest things to make,' Silas says. 'I have enough trouble mowing stripes on my lawn.'

'The academic community is taking this very seriously,' Strover continues, ignoring Silas's attempt at humour. 'Sees it as a challenge.'

'Aliens testing our IQ, if you believe the stuff online,' Silas says, rolling his eyes. 'Apparently, they need to know that we have sufficient intelligence before deigning to communicate with us.'

Strover's not playing ball today. Not in the mood for *X Files* banter. 'Have you heard anything from MOD Police at Porton?' she asks.

Silas shakes his head. He hasn't even rung them yet. There are other more important lines of inquiry to pursue. The key has yielded no forensic clues. Silas woke up this morning still convinced that the third victim holds the secret to the other two. It's just a question of establishing his identity. No match came back from the National DNA Database and forensics have yet to discover anything useful at the scene.

'What about the female victim's tattoos?' Silas asks. A small rook has also been discovered behind the victim's left ear, hidden beneath her hair.

'Seems like the tattoos on her arms are rooks too,' Strover says. 'Wing feathers.'

Nature can be a cruel bastard. It was rooks that had gathered in the field to peck at her body. Feather tattoos are popular but someone must know who this woman is. They've drawn a complete blank on her and the man found earlier.

Silas's phone rings. It's Malcolm, the pathologist. Silas left him a long message this morning about their latest discovery.

'I've just come off the phone to the consultant at the Great Western looking after your "zombie", as you call him,' Malcolm says on speakerphone, clearly disdainful of the term. He's never one for small talk, likes to get to the point, which suits Silas fine. 'Actually, it got me thinking. I told him to test for tetrodotoxin.'

'What's that when it's at home?' Silas asks, glancing at Strover, who immediately pulls out her phone. She likes to look up anything that she doesn't know.

'An extremely dangerous neurotoxin – found in puffer fish and more deadly than cyanide,' Malcolm says. 'One of the most potent non-protein poisons known to man, in fact. For many years, it was thought to be an ingredient of a Haitian voodoo preparation used to induce a state of living death – zombieism. The science is dubious, to put it mildly, and I suspect your man was poisoned with something else, but he's apparently developed fixed dilated pupils and brain stem areflexia – the cardinal hallmarks of brain death.

Tetrodotoxin is also often administered through a cut or wound.'

'Thanks for that,' Silas says, recalling the slash on the man's arm. He glances across at the party of French detectives. One of them has turned around at the mention of Haitian voodoo.

'The victim's most probably right-handed – there's a watch mark on his left wrist – so there's a chance the wound, on the left forearm—'

'Could have been self-inflicted?' Silas interrupts.

'Exactly,' Malcolm says, pausing. 'Unlike the gash on his cheek. Most likely a deer. I've run it past an American colleague who works at one of the body farms over there, as we don't have many cases of flesh-eating Bambis in the UK.'

Silas knows all about the body farms. Run by forensic anthropologists, they are places where human remains can be studied as they decompose in an outdoor environment. The UK uses dead pigs but human corpses are permitted in the States, scattered around high-security grounds. A few years back, the Forensic Anthropology Research Facility in Texas made the news when a white-tailed deer, considered a herbivore, was photographed chewing on a human rib.

'You said in your message that you think your zombie might have been responsible for the other two bodies,' Malcolm continues.

'It's just a hunch,' Silas says.

'An interesting one.' Malcolm pauses. 'It's almost as if whoever did this was an insider – sending a message from one pathologist to another.'

'How do you mean?' Silas asks.

'The first victim's brain had been severed after death, and

both bodies had been frozen. You could argue that's our stock-in-trade, what we do.' He pauses. 'It's all a bit cryptic, isn't it? What with the key and everything.'

'Can say that again,' Silas says, glancing at Strover, who has taken a call on the other line.

'More doosra than off break,' Malcolm adds, using one of his favourite cricket analogies. If it wasn't for Conor, who's back playing for their local cricket club, Silas wouldn't know what Malcolm is on about.

He moves to hang up but Malcolm, usually so brusque, stays on the line.

'You can imagine there's been a fair amount of chat among us pathologists,' he says. 'We're a small community, overworked, underpaid. As a breed, we're possibly even a little arrogant, but we're not monsters. Whoever did this might well be a pathologist but they had no respect for the dead.'

'I'll keep you posted, Malcolm,' Silas says, looking again at Strover, who is gesturing at him. 'Got to go.'

He hangs up and turns to Strover, who has her hand over the receiver.

'Sir, it's the Control Room. Reports of an RTC coming in near Middle Wallop – involving a Range Rover. Same number plate. Two casualties. They're being brought in to the Great Western now.'

36

Jim

Jim knows it's a risk but he needs to see his dad. He can't help feeling that it might be the last time they meet. It can take him more than two hours to drive to Swanage on the Dorset coast, but that's on a Friday night and right now the roads are clear. His body is still running on adrenaline after the incident with the tractor as he heads back towards Marlborough. It's in the opposite direction, but there's somewhere he needs to visit before he heads south to Swanage.

Jim glances in the mirror, checking for another tail. The same car has been behind him ever since he doubled back onto the Salisbury road after leaving Palestine. It's just a mother and daughter. No cause for alarm. He hopes no one died in the Range Rover, but they pushed him too far, asking him to pull over like that. They were definitely MI5, tipped off by MOD Police at Porton Down. DI Hart will have contacted them about his meeting with Bella at the pub last night – it was the first thing the Range Rover driver asked him about on the phone – and all hell will have let loose. Calls will have been made between various Whitehall departments, checking Jim's Developed Vetting security clearance, his recent computer activity. They won't find anything.

Thirty minutes later, he pulls into a lay-by at the top of Hackpen Hill, outside Marlborough, steps out of his car, and walks over to the brow. To his right, a spinney on the ancient Ridgeway trail. Ahead, towards Basset Down, the distinctive shape of the Science Museum's National Collections Centre at Wroughton, two large Nissen huts where spacecraft, aeroplanes and vehicles are stored. Dad used to take him there for visits when he was a child. An electric car from 1916 was one of his favourites. And below him, a spread of wheat fields, the closest of which is humming with police activity. Tents, CSI officers in white oversuits, a gathering of camper vans in the distance. But Jim's not come here to gawp, unlike the prurient crowd that has assembled in the car park opposite. He's here to study the pattern, the representation in flattened wheat of the incapacitating agent BZ: 3-Quinuclidinyl benzilate.

He's looked at the photos but he wants to see it for himself. Check that he's right, particularly after what happened with the Range Rover. Shielding his eyes from the sun, he takes in the geometric shapes and a moment of doubt passes through him like a breath of autumn wind. Has he got the molecular structure wrong? Misunderstood the message? Is it something else other than BZ? He squints, looking at the two hexagons and the less distinct symbols beyond them. If it's not BZ, it's something very like it.

A second later, another Range Rover, black and sinister, similar to the one he'd caused to crash, noses around the tight bend below and drives up towards him, slowing as it passes. Jim turns away, hiding his face. Is he just being paranoid?

Jim gets back into his car as the Range Rover accelerates away towards Marlborough. He starts to text Bella, but his hands are shaking too much. If she's still sorting Rocky in his house, she needs to get away from there now. He doesn't want her hurt by these people.

37

Bella

Bella finds the key under the flowerpot at the back of Jim's cottage and unlocks the door. As far as she knows, no one saw her walking down the drive and she checks both ways before she steps inside. It's strange being back here again. A risk too. Should she really not trust the police, as Jim says? Bella's childhood in Homerton was played out against a soundtrack of wailing sirens but she was too focused on her schoolwork to get into trouble. Besides, the local police were more interested in the rival gangs of Tottenham and Wood Green, although her mum occasionally accused them of being heavy-handed with some of the refugees at the migrant centre.

Bella looks into the sitting room. Everything's been tidied to within an inch of its life. Ordered, how she imagines Jim to live. It's as if last night never happened. There's a row of framed photos on the mantelpiece: a younger Jim in his graduation gown; by the seaside with an elderly man, presumably his dad. And a photo of Jim standing in front of a sign saying 'Porton Down Science Campus'. Next to it, on the right, are the logos for DSTL, Public Health England and Porton Science Park.

She glances over at Rocky in his leafy vivarium. The chameleon eyes her with suspicion as she walks across to peer at him through the glass. He's a weird-looking boy: pastel green with mottled stripes and a strange cone-shaped skull, like an egghead. Maybe he's a brainy scientist, like Jim. Lifting off the lid, she looks down at Rocky and hesitates. Animals don't scare her. They had all sorts of pets in Mombasa, including Cadogo, their Alsatian guard dog, and Kasuku, a vicious African grey parrot. And in London one of her neighbours used to keep a ball python, which Helen once dared her to hold. Reaching down, Bella lifts up the stone, removes the USB stick and gives Rocky a little wave.

Should she have a quick look at the USB now? Jim said it was encoded but something makes her want to try. She's already checked out of the pub, confirming that they were happy to have Rocky, and she has a small bag with her. Sitting down in the armchair, she pulls out her laptop and inserts the USB. It's called 'Modern Maddison' but when she clicks on it, a password window appears. What happens if she never hears from Jim again? He should have trusted her, given her the password.

Bella's phone buzzes with a text. It's from Jim.

If you are in my house, you need to get out of there now. Forget all I said and go.

She stares at the message and reads it again, her mouth drying. Has she triggered a warning by trying to open the

USB? She looks around the tidy room. The front doorbell rings and she freezes. Jim said he'd been followed. The Range Rover must be back in the village. Or is it the police? She glances again at her phone. Something about the tone of Jim's text is not right. Should she leave for her own safety, or does he mean that she should get out of his life? She closes her eyes, tries to calm down. They got on well last night. He's told her how to enter his home, even where the USB is. After one meeting. He trusts her.

She walks over to the door. It's probably someone from the pub, offering to help bring Rocky over. Except that they agreed she could manage on her own. The blurred outline of a figure is visible through the opaque glass panel. The bell rings again. She decides not to answer it and the person walks away. Running through the house, Bella slides the key into the lock of the back door, turning it silently. A moment later, the person reaches the door and tries the handle. Do they have their own key?

She holds her breath and listens, tensing every muscle in her body. The person raps their knuckles so hard on the door's glass panel that Bella fears it will break. Is it the same person who struck Jim on the head last night? Her body tenses again. Shit, shit, shit! What's Jim got himself into? What's she got herself into? She tries to keep quiet, rechecking her phone. Should she text Jim back, tell him someone's here already? Ask him what he means?

The person starts to walk around to the front of the house again. Bella tiptoes upstairs to the bathroom, as quietly as she can, closes the loo seat and stands on it. She can just see down to the drive outside. A man is walking away, about to disappear around the corner, but not before Bella recognises his hunched shoulders and jet-black hair. The same man she confronted in the street outside the newspaper office. Who stepped out of the car near her house in Homerton.

What's he doing here? She waits until he is gone, breathing hard, trying to stop a rising panic, before she returns downstairs. Why is this man following her? She needs to get out of this village, back to London, the safety of home, her mum. Perhaps she can sneak down to the train station without being spotted. She looks again at Rocky. He needs to get to the pub. She'll call them, ask someone to come over and collect him, say the vivarium's too heavy.

She jumps as Jim's landline rings. Is it Jim? Maybe the man who just came to the door? After five rings, the landline goes to answerphone.

'Hi, Jim, just checking to see if you're coming in to work today?' The female voice is local, cheery, trying to sound casual. 'We're a bit worried you haven't shown up yet. Give us a call if there's a problem. Thanks.'

Bella waits for the answerphone to switch off. Was that Porton Down calling him? It didn't exactly sound like a secret government facility. There were voices in the background, almost as if the woman was calling from a shop. But what does she know? She goes over to the phone and dials 1471 to see if there's a record of an incoming number. To her surprise, there is and she takes a note of it, calling the number on her mobile.

'Hello?' The same breezy voice. 'Porton Garden Aquatic and Pets, how may I help?'

38

Silas

'What do you mean, they've discharged themselves?' Silas asks.

'Welcome to patient autonomy,' the doctor says. Silas is with the senior registrar in charge of A & E at the Great Western Hospital in Swindon and the doctor is in no mood to chat. Patients are on trolleys queued up along the corridor behind him and earlier Silas saw ambulance crews in a line outside, waiting to sign off their human cargo.

'How long ago was this?' Silas asks, still despairing as he takes in the crowded scene. He doesn't blame the doctor. He should have come straight here, met the two occupants of the Range Rover as they arrived by ambulance, but Ward wanted a quick chat to find out how the investigation was going.

'About fifteen minutes ago,' the doctor says. 'If you'll excuse me...'

'Of course,' Silas says, glancing at Strover, who stands back to make way for an elderly patient being pushed past in a wheelchair. 'Did they sign anything?'

'We got them both to sign the usual legal forms to say that they were leaving against medical advice and that we could no longer be held responsible for their health.'

Forms mean signatures, which mean names. Probably false ones, but at least it's something.

'Can I see them?' Silas asks. 'The forms?'

The doctor shakes his head in disbelief.

'Take a look around, inspector,' he says. 'Does it look like I've got time to dig out the paperwork for a couple of people who suffered whiplash, maybe mild concussion, and some minor cuts? They weren't even meant to be here – the ambulance should have taken them to Salisbury District.'

'Why didn't it?' Silas asks. He'd forgotten that the hospital in Salisbury was nearer to the accident site than Swindon.

'I've no idea,' the doctor says, exasperated. 'And at this point I really don't care. That patient over there' – he gestures at an elderly woman – 'is eighty-seven years old. She had a fall at home and has been on a trolley for thirteen hours.'

Silas has nothing but admiration and respect for the staff here. It's where Mel used to work as a nurse. They also looked after his father in his last days with care and love that was beyond the call of duty. But he needs to find out everything he can about the occupants of the Range Rover that crashed today. It's no coincidence that the accident occurred so close to Porton Down. There are also reports from a local farmer that another vehicle was responsible for the incident and drove off at speed.

'I appreciate that,' Silas says. 'Maybe we could return at a quieter time.'

The doctor laughs. 'Let me know when that is – I must remember to come in to work that day. If I haven't retired by then. Or died.' He pauses. 'Give me your number and I'll see what I can do.'

'Thanks,' Silas says, glancing around him again. These are good people working here.

Ten minutes later, Silas and Strover are at the entrance to the intensive care ward, waiting to see the man they found alive last night in the crop circle at Stanton St Bernard. Silas feels like a doctor on his rounds. He's tried to visit Noah but he's still not fully conscious. Maybe later.

A nurse ushers them into a small family room off the main ward. Was it in here that Silas sat down with a consultant and nurse to discuss turning off his father's life support? This room or the one next door. Those days are a bit of a blur.

'Two teas,' the nurse says, placing a couple of mugs in front of them. 'One with, one without. Biscuits are in the tin.'

'Thanks,' Silas says, glancing at Strover and then the tin. He's going to have a biscuit even if she isn't. It might be his only chance of decent food today. He'd dropped off some wine and shortcake after his father had died – a thank you for the care the staff had shown him – so it's not all one way. At least, that's what he tells himself as he tucks into a chocolate digestive.

'He's a popular man today,' the nurse says.

'Who?' Silas asks, flashing a look of concern at the nurse.

The nurse nods in the direction of the ward. She knows they're police, rather than relatives, and her manner is more gossipy than compassionate.

'Bit of a medical curiosity,' the nurse continues. 'We've had them all in – consultants, surgeons, the lot of them.'

Silas sighs with relief. In his chat with the boss, he had floated the idea of asking for an officer to stand guard at the door, in case the man became a target, but it hadn't gone down well. Not a good use of resources. Hospitals are secure places these days, cameras everywhere.

'The doctor's finished, you can go in now,' the nurse says, watching as a man in a white coat walks past her and down the corridor. 'Don't recognise half the people who work here these days,' she adds.

Silas tenses again. He'd clocked the doctor too. Was there something a little too rushed about him, even for a busy medic? A hint of overstretch in his stride?

39

Bella

Bella sits back on the train with a sigh of relief, pleased to be out of the village and on her way up to London. Her carriage is busy but at least she's got a seat. By the window too. For the first time she sees the canal, running parallel with the train line. She should have gone down there for a walk, enjoyed the country air, but she hasn't had a moment. The last hour has been frantic.

She'd left Rocky and his vivarium on the back doorstep of Jim's house, after ringing the pub to say that she was running late for the train. She'd even remembered to leave out a pot of live locusts, after the landlord had asked about Rocky's food. Checking the timetable, she'd chosen her moment carefully, rushing down from the house to arrive at the station just as the train was about to leave. No one saw her, as far as she knows, and she didn't see the Range Rover or the man who had come to the door.

She still can't work out the message on Jim's answerphone. According to Google, Porton Garden Aquatic and Pets is around the corner from the secret government site. *Just checking to see if you're coming in to work today.* She assumes Jim visits the shop – for locusts – during his lunch break at work or before he starts his working day. That text message is still troubling her too. *If you are in my house, you need to get out of there now. Forget all I said and go.*

She pulls out her phone. She hasn't texted him back yet – she wanted to get her head around everything that's happened – but she decides to send him a message, tell him she's left the house and sorted Rocky. But then she overhears the conversation starting up further down the carriage and her blood runs cold.

'Apparently her arms were covered in feathers,' a woman is saying. 'And she had a rook right behind her ear. Spooky or what.'

'I wouldn't like that,' another woman says. 'If I got a tattoo, I'd want a pretty little rose on my arse.'

'I wouldn't want my dead body left in a crop circle overnight, thank you very much.'

Bella wants to be sick. The carriage starts to spin and she steadies herself against the window with a hand.

40

Silas

The nurse pulls back the blue curtain and lets them stand by the man's bed in the intensive care unit. Silas is shocked all over again by his appearance. The zombie victim's skin is still a sickly white, as pale as pallor mortis, the waxy, telltale colour of recent death. Except that he's alive. Head tilted back, he's on a ventilator, a tube running into his nose, his wrist connected to a drip. The only obvious sign of life is the heart monitor behind him, a green pulse tracking across the screen. Silas is not sure why he's here. He needs to reassure himself that the man is still of this world. And he doesn't want last night's image of him to be his last. The scene before him is almost worse.

'There is one thing that might be helpful,' the nurse says.

'Go on,' Silas says. He could do with some good news.

'He stirred in the night,' the nurse says, glancing down at the patient.

'Stirred?' The man looks as if he hasn't moved for years.

The nurse nods. 'I came over straight away. He was trying to say something but I couldn't hear what it was.'

'He tried to speak?' Silas asks, checking with Strover, who is equally surprised.

'I leant right in close to him but I couldn't make out what he was mumbling. Sounded French to me. So I called over Françoise – she's from Marseilles,' the nurse adds, pronouncing the town like 'Ma says'.

'Is she here?' Silas asks, desperately hoping this isn't going to lead to nowhere.

'Her shift's over but she thinks she knows what she heard.'

'And?' Silas asks, failing to conceal his impatience.

'My French isn't great but he might have said *"folie à deux"*.'

Silas looks at Strover again. 'Madness of two?' she offers. That's what he thought. Give or take.

'Thanks,' Silas says to the nurse, making a note in his phone. 'That's very helpful.'

'I'll leave you to it, then,' the nurse says. '*Au revoir*,' she adds, stepping outside the curtain.

'*Folie à deux*,' Silas repeats, trying not to butcher the French language any further. 'What the hell's that about?'

'Maybe there's a connection between the first two victims?' Strover offers. Silas nods. He's already told her his theory that the man lying in front of them was responsible for placing the two dead bodies in the first two crop circles.

'They were both young, could have been lovers, I suppose,' Silas says. It doesn't sound right and they both know it.

'What about the blood tests?' Strover asks.

'Should be back shortly,' Silas replies, still mulling over the implications of *folie à deux*. He's amazed that the man before them could even speak.

'Think they'll find any tetrodotoxin?' Strover asks.

'If I was a betting man, I'd put my house on it.' As it happens, Silas has become a betting man, but Strover doesn't need to know about his recent flutter on the horses.

A moment later an alarm goes off on one of the monitors behind the patient.

'What's that about?' Silas asks. An image of his father on a ventilator comes and goes. The alarm had sounded for him too.

'His oxygen levels,' Strover says, looking at the bank of monitors. 'Dropping like a stone.'

The nurse rushes back in, checks a reading on one of the screens and calls for assistance, an unmistakeable sense of urgency in her voice.

'I must ask you to move away, please, into the corridor,' she says.

'What's wrong with him?' Silas asks, stepping back from the patient. They hadn't been able to save his dad but his passing had been a mercy in the end.

'I don't know yet,' she says, adjusting a dial. 'Please, we need space.'

Silas and Strover step further away as two other medical staff join the nurse around the bed. And then they are wheeling the patient out of the ward through swinging doors, the air filled with shouted instructions and orders.

'Nurse, who was that last doctor who saw him?' Silas says, as she rushes back into the ward to collect a clipboard.

'No idea,' she says. 'Why?'

'Because I don't think he was a doctor.'

41

Bella

Erin's dead. Bella stares ahead in the train carriage, stunned, trying to come to terms with what she's overheard. The second victim had a rook tattoo behind her ear – just like Erin. How's her best friend gone from being ill in hospital to her dead body being dumped in the middle of a crop circle in Wiltshire? It can't be true, makes no sense at all.

'You alright?' a young woman sitting opposite asks. She picks up Bella's phone and hands it to her. Bella hadn't even realised that it had slipped off her lap onto the carriage floor.

'Thank you, I'm fine,' Bella manages to say. But she's not. Not fucking fine at all. She can't even begin to get her head around what's happened to Erin. How can it be? She was meant to be seriously ill in hospital. Should Bella have done more? Demanded to see her? She bites her lip and stares out of the train window.

'You sure you're OK?' the woman asks.

But Bella doesn't hear. She's back in Oxford, Freshers'

Week. *Bel! Bel! They're fucking killing me! Help me!* Erin was screaming for her, but Bella could do nothing to stop them pinning her friend down on the ground, as if she were a wild animal, forcing the drugs into her. There were too many people, off their heads. Bella tried to intervene – God, how she tried – but they pushed her away, throwing her to the floor. Another hideous initiation rite that had gone way too far. And Erin was always in the thick of it, drawn to trouble like a moth to the flame.

'Hazing – it was student hazing,' Bella says to the woman opposite. 'You know, like in America.'

The train carriage starts to spin and she closes her eyes. At least Erin survived, didn't blame Bella afterwards for not doing more. She was that kind of friend. Forgiving. This time she's dead. And it's all Bella's fault.

'Shall I call someone for you?' the woman asks, a look of growing concern on her face.

'I'm getting off at the next station,' Bella says, as Erin's despairing cries start to fade. 'Just need some fresh air.'

'Let me help you, then,' the kind woman says, watching Bella rise unsteadily to her feet. 'We're here now.'

A minute later, Bella is standing alone on the platform as the train pulls away. She'd had no intention of disembarking at Hungerford but she couldn't stay on board a moment longer. The woman had checked she was OK before getting back on the train herself. Bella sits down on a metal bench in a daze and pulls out her laptop. There's a weak free Wi-Fi signal from the pub across the car park. Her hands shake as she searches for 'crop circles' and 'feather tattoo'. *Christ, please don't let it be true.* But it is.

Mystery of naked 'bird woman' found dead in crop circle.

It's too much. She spins away from the headline and throws up. Tears stream down her face as she's sick again. Gradually, her retches turn to sobbing gulps. She looks around. The station is empty, thank God. She starts to scroll through the text. She can't do this. Glancing up at the sky, she takes a deep breath and closes her eyes. She needs to find out more – for Erin – and begins to read again.

No one knows who the woman is and police are appealing for any information about her and the unusual tattoos. Erin never showed anyone the rook behind her ear, apart from Bella. It was their secret, shared late one night near the end of Bella's final year, when Erin was in her room. She'd got the tattoo as a child. Forbidden, hidden. That evening was also the only time Erin danced, reliving the mesmerising Irish reels and jigs she used to perform as a child. Afterwards, she tried to teach Bella the treble, but her long legs got into a terrible tangle.

'You need a rook behind your ear, you feckin' skinny malink,' Erin had joked. 'It calls out the moves – one, two, three! One, two, three!'

Bella continues with the article, blinking away the tears, trying to focus. She can't believe what she's reading. The police suspect that Erin was murdered, possibly a victim of a sex game that went wrong. Her body was found naked in a straitjacket. Bella shakes her head, banishing the image. Erin wasn't into that sort of stuff, not as far as Bella knew. At the end of the article, she recognises the name of the Swindon CID detective who is quoted. DI Silas Hart – the man who came to Jim's house. Checking the number that's given at the end of the story, Bella dials it and waits.

'Hello?' she says, glancing up and down the platform.

She takes a deep breath, trying to stop the sobs. 'Is that Crimestoppers?'

'Yes it is, how can we help?' says a woman's voice. She sounds calm, reassuring, but Bella is still nervous.

'I know who the crop circle woman is,' she begins, sniffing, wiping her nose with the back of her hand. 'The one whose arms are covered in feather tattoos. In Wiltshire. She's a friend of mine.' Bella starts to sob again. 'My best friend.'

'OK,' the voice at the other end says. 'I'm so sorry to hear that you've lost someone close to you. Would you like to give me her name?'

'Erin, she's called Erin,' Bella says, looking around her again.

'Do you have a surname?'

Bella is about to continue when she hears Jim's words ringing in her ear. *Do not trust the police if they get in touch again, do you understand?* Crimestoppers is not the police. It's an independent charity, isn't it?

'Hello? Are you still there, caller?' the woman asks.

Bella moves the phone away from her ear and stares at it, listening to the voice, its gentle but repeated requests for her to continue.

Remember, the police are not your friend.

She ends the call and rips out her phone battery as if she's gutting a fish.

42

Silas

'I want the footage from every bloody camera in this place,' Silas says, storming into the office of the director of estates at the Great Western. Strover is behind him, hurrying to keep up with her boss. He has just asked for a full lockdown of the hospital and called for as much uniform backup as nearby Gablecross can muster. The helicopter should also be overhead within the next few minutes.

The 'zombie' patient, their main suspect for the crop circle killings, died shortly after going into cardiac arrest and Silas is convinced that the 'doctor' they saw walking out of the intensive care unit was responsible. Silas should have acted on his original suspicions. After the patient's condition had started to deteriorate, he had rushed out of ICU with Strover in an effort to find the fake doctor, but their search had proved fruitless in the hospital's rabbit warren of corridors.

The director is now doing all he can to reassure Silas as he paces around his small office, reminding him that the hospital has always worked closely with Wiltshire Police. It also has a team of security guards on duty 24/7 and strong surveillance systems across the estate.

'If he's still on site, we'll find him,' the director says. 'If not, we'll have a good photo.'

'I hope you're right,' Silas says, turning to Strover. 'We also need a twenty-four-hour police guard outside Noah's room,' he says. At the moment, it's being protected by one of the hospital's security guards.

'We run very rigorous checks on all our staff,' the director says, sitting down in front of a computer at his desk with a reassuring sense of urgency. 'He would have needed a security card to gain access to ICU – and we keep records of everyone who swipes in and out of every secure part of the hospital.' He starts to type and a window of data opens up on his screen. Silas watches as his whole demeanour changes. 'That doesn't make sense,' he says.

'What doesn't?' Silas asks, glancing across at Strover.

'Doesn't make any sense at all,' the director continues, to himself as much as to the others. 'According to the security data, the consultant who checked on the patient was Dr Armitage. He's one of our most respected doctors. There's no way—'

'And one of the most recognisable?' Silas asks. 'His face would be familiar to nurses?'

'Of course, everyone knows Dr Armitage, why?'

'Because the ICU nurse didn't recognise him when he walked past.'

'Dear God,' the manager says, putting his head in his hands.

Silas closes his eyes. He'd feared as much. The director

double-checks the data on his screen and then Strover's phone rings.

'That was the A & E registrar,' she says to Silas, after taking the brief call. The one who hadn't warmed to his brusque manner earlier. 'He's dug out the paperwork for the two people in the Range Rover who discharged themselves.'

'And?' Silas asks, impressed. Strover's charm must have got her further with the registrar than he'd managed.

'He says there's a problem,' Strover says, turning to the director. 'Seems like the consultant who signed off their papers was the same Dr Armitage – except that it wasn't his signature. Nothing like it, apparently. They'd also insisted that the ambulance took them to Swindon not Salisbury after the crash. The paramedics agreed, but it was an unusual request.'

Silas nods approval at Strover. She's done well, established a connection between the fake doctor and the two people in the Range Rover. They're not dealing with a few individuals here. It's a well-organised network of people intent on shutting down anyone capable of spreading light on the crop circle deaths. He just wishes he knew why.

His own phone starts to ring as they leave the director's office and head for the car park. It's the Control Room.

'Sir, we've just had a call from Crimestoppers – for the crop circle case.'

Silas stops in the corridor as a group of uniformed police officers come out of the lift. It's about time they had a break.

'Go on,' he says, trying to disguise the eagerness in his voice as he signals for the uniforms to wait.

'A member of the public rang up, said they knew the second victim, the female.'

'Have we got a name?' Silas asks. 'For the victim?' More officers stream down the corridor. It's been a good response from uniform for once.

'Erin. Echo. Romeo. India. November. No surname.'

'And the caller's name?' Silas asks, more in hope than expectation. Crimestoppers makes a point of never asking for a name or the number of callers and conversations are not recorded. They are also scrambled to prevent the identification of numbers and callers being rung back.

'I'm sorry, sir. No name for the caller. Just said the victim was a friend.'

43

Jim

Jim brings the car to a halt by the tennis courts above the beach in Swanage, opposite the house where he grew up. It's empty now, been on the market for more than a year. The clay soil below the house dried up during a hot summer a few years back, creating cobwebs of alarming cracks inside. He'll come back later, after he's been to visit his dad, who now lives in a small modern apartment on the far side of the bay. It's a ten-minute walk, a chance to order his thoughts after the drive. No one seems to have pursued him but he feels happier not leaving the car outside his dad's new place, just in case.

One thing bothers him, though, as he cuts down to the beach and makes his way along the busy seafront. Why is the person who appears to be sending him messages doing it so publicly? And so morbidly? As far as he can tell, none of the crop circle victims has been identified yet, but Jim's certain they will be connected in some way to Porton Down – victims, perhaps, of ongoing experiments at The Lab. In which case, the person must know what

Jim's up to, the evidence he's been collating since returning from Harwell.

He walks on, scanning the holidaymakers as they spill out of the amusement arcade onto the crowded beach. No one he recognises. Swanage is in full summer holiday mode, the air filled with the smell of fried food and sun cream. The bay is buzzing with jet skis and sailing boats, the horizon littered with empty cruise ships, a legacy of Covid-19. To Jim's right, above the seafront, a nodding T-Rex looks down on the holidaymakers. Jim stops to look and smiles. The scene is more surreal than usual, the dinosaur's animated jaws seemingly singing along to a sea shanty blasting out from an adjacent pirate-themed golf park. *Fifteen men on a dead man's chest, Yo ho ho and a bottle of rum...*

He starts to relax. No one's followed him here. To his left, the beach is punctuated by a row of wooden groynes to prevent longshore drift – a local phenomenon that must have been studied by almost every geography field trip student in the country. The Punch and Judy tent draws people from far and wide too. Young children sit cross-legged on the sand, laughing as Punch attacks a crocodile with his slapstick. Behind them, families eat fish and chips on the wall as joggers zigzag through the crowds on the promenade.

It could have been a sad life but Jim was happy growing up here. An only child, he was self-contained, scouring the local shoreline for fossils, finding meaning in their shapes and patterns. The geometry of the Jurassic coast. And then there were his reptiles, a source of much joy at home – and of bullying at school, until he was big enough to look after himself. It was university life that he found more challenging.

Jim turns up from the waterfront and sees the block of apartments ahead. His dad downsized into one of them a year ago, just before his health started to deteriorate. It feels strange coming to his new home, an unfamiliar place in a town he knows so well. After the carer lets him in, he finds his dad staring out to sea in the front room. Tiny specks of dust circulate in the bright sunshine, moving around him like agitated atoms. He doesn't turn when Jim enters the room, but that's not unusual. He's got dementia and has good days and bad days. Today is a bad day, according to his carer.

'Hi, Dad,' Jim says, as he kneels down in front of him and takes his wrinkled hand. His hair is thin and snow white and his whole body seems to have shrunk, as if it's been washed at the wrong temperature. There's no change in his dad's expression, just a distant smile out to sea. Jim glances over at the carer for reassurance as his eyes well up. He's used to seeing him in this state but it still shocks him every time.

'How have you been?' Jim asks, more in hope than expectation. No response. 'The bay is busy but no sign of those nasty offshore powerboats you don't like.' Jim's putting it mildly. Before his health deteriorated, his dad wrote angry letters to *The Times*, complaining about the noise pollution.

'Do you want me to play the piano for you?' Jim continues. Out of the corner of his eye, he sees the carer nod enthusiastically at the door, arms folded across her formidable chest.

'OK then,' Jim says. 'As it's you.'

44

Bella

It's an hour before the next train to London and Bella feels vulnerable on the deserted platform at Hungerford. Confused too. How did Erin, the one person she could be herself with, who really understood her, end up dead on a hillside in Wiltshire? Why didn't she do more, go to Oxford, demand to speak to Dr Haslam? Sometimes it feels as if she's looking at her own life through a window, a witness rather than a participant. Not fully engaged.

She glances around the station. Could it be her next? She's never been to Wiltshire before and the first time she visits, her best friend's body is found in a nearby field. Who sent her that anonymous typed letter? She should have pushed Jim further about the connection between Porton Down and the crop circles, asked him who might have wanted them to meet in the pub.

Her hands are still shaking as she walks away from the station. It was a mistake to ring Crimestoppers. What if they do pass on her number to the police? They could establish her identity and location using the mobile phone network.

Maybe she's mistaken about Erin? There's no proof it was her. Lots of people have bird tattoos. But she knows with sickening certainty that it's Erin. No one else would have a tiny rook behind their ear. She just doesn't understand how she came to be murdered.

Perhaps Erin was better than everyone thought and discharged herself from hospital? Dr Haslam stopped returning Bella's calls because he was embarrassed. He didn't know where Erin was and had failed in his duty of care. In which case, why didn't Erin ring her as soon as she was out of hospital, reply to all her messages? Maybe her murderer met her in hospital? Or did she become desperate and take her own life? So many unanswered questions. All she knows is that her best friend was too young to die. Too funny. Too crazy.

She looks for a payphone on the high street in Hungerford and finds one outside a bookshop. It's not safe to make calls on her mobile. She's also unusually familiar with a payphone for someone in their twenties, as her old mobile is usually out of credit or charge. She calls her mum, a number she knows by heart, and inserts a pound coin. Annoyingly, it goes straight to answerphone.

'Mum, it's me,' Bella says, glancing up and down the street again. 'I need to talk. Can you leave a message on my mobile? I've had to turn it off but I'll check it in a bit.' She hesitates, wondering whether to say any more, suddenly overwhelmed by the news of her friend. 'Please call me, Mum,' she says, starting to cry. 'I think Erin's dead.'

Saying the words out loud shocks her all over again. She searches her bag, tears smudging her make-up, and finds another pound coin. It's unlike her mum not to answer. At college, Bella might not have been in touch as much as she should, but she knew that she could ring her mum at any time of the day or night. That was the deal. She tries the landline and gets another answerphone. 'Please, Mum, can you call me. Erin's dead. Dr Haslam's been lying to us. She wasn't in hospital. Someone's killed her. Where are you? I'm coming home.'

Bella backs away from the phone box, desperate to speak with her mum. She needs to get a grip, stay sharp. Checking around her, she's about to step into the bookshop when she spots a police car at the top of the high street. Could the police have already established that she made the call from Hungerford train station? Bella retreats into the doorway of the bookshop like a hunted fox, eyes fixed on the police car, as it turns up the side road towards the station.

They know.

45

Jim

Jim walks over to the small piano in the corner of the front room and opens up the lid.

'Are you ready?' he asks, glancing across at his dad, who's still staring out to sea. He follows his gaze. Can he see the chalk-white cliffs of the Isle of Wight in the distance, rising above the horizon like a giant breaking wave?

His dad used to play well himself and loves his music. It's only when he stopped going to hear the Bournemouth Symphony Orchestra on a Wednesday night in Poole that Jim knew his illness was serious. The annual series of concerts were an immutable part of Jim's childhood. Stopping off for chips on the drive back afterwards, discussing the acoustics of the hall. After years of experimentation, his dad had settled on seat 27, row M as the best place to listen.

'Then I'll begin,' Jim says.

For a moment, he thinks of playing one of the 'Goldberg Variations', but he knows what works best. The instrumental version of 'Bring Me Sunshine'. On good days, his dad joins in, but he doubts he will today.

Slowly, he begins to play, stroking the keys with affection, fighting back the tears. And then he hears his dad start to sing along, faintly at first and then more confidently. Jim speeds up and he responds. Morecambe and Wise's version used to drift up from downstairs when Jim was younger – along with his dad's singing and laughing.

After he's finished, Jim returns to his dad's side and holds his hand again. This time he pats Jim's fingers, a faint ripple of recognition, of affection, and turns to look at Jim with his tired, watery eyes. He's aged again since Jim was last here, barely a month ago.

'I've got to go away, Dad,' Jim says. 'I'm in trouble at work but it's for a good cause. I just want you to know that.'

'You always had a strong sense of right and wrong,' his dad says, catching Jim off guard with the sudden clarity of his response. Jim glances over at the carer, who has remained at the door.

'You've woken him up,' she says, smiling.

'Looks like it.' Jim pauses. 'Can we have a minute? On our own?'

'Of course, sorry,' she says, retreating to the kitchen. 'Let me know if you need anything.'

Jim turns again to his dad. 'I need to tell the world what's going on at The Lab. I was beginning to lose my nerve but someone on the inside has started to encourage me. Send me signs. I've got to do the right thing, Dad. You always taught me to do that, didn't you?'

His dad nods, his eyes distant again. He used to be a

lecturer at Southampton University's department of engineering, specialising in acoustics. His mum, who had a history of mental illness, took her own life when Jim was two, leaving his dad to bring him up on his own. A tough call for anyone, but particularly for an older man. His dad had just celebrated his sixtieth birthday.

'How's work?' his dad asks.

'It's good,' Jim says, turning away. He's lost him again. A tall ship cuts across the bay, its white canvas sails iridescent in the summer sunshine. A cargo ship from Jersey heads into Poole in the other direction.

'I still blame myself – for what happened,' his dad says.

'Don't be silly,' Jim says, patting his hand, pleased to have him back. His skin is like tracing paper, thin and transparent. 'It wasn't working out with my team leader at Porton so I put in for a secondment and got to do some amazing science at Harwell. It was tough at times but now I'm back at Porton and it's all good.'

'She was so pleased when you were born,' his dad continues. 'I thought it would change everything.'

Jim realises that he's talking about his mum and her mental illness and not his own problems at work. He looks out of the window again and his heart nearly stops. A black Range Rover is crawling along the crowded seafront, towards the Pier car park, as if the driver is looking for someone. The bastards. He was being followed.

'I have to go,' Jim says, getting up from the chair, eyes still locked on the Range Rover.

'So soon?' his dad asks. Sometimes Jim wonders if he's happier in his own world. When music brings him back to the present one, as it seems to do, his life is only open to disappointment.

'I love you, Dad,' Jim says, leaning down to kiss the top of his head. 'I'll come and visit you again soon, when all this is over. I promise.'

46

Bella

Bella hovers in the Hungerford Bookshop, trying to concentrate on the books laid out on the table. She picks one up, puts it down. Then another. Normally she would be in heaven. When Bella was younger, her mum used to take her to the library in Hackney, where she'd spend hours sitting on the floor reading children's books. Her dad used to read to her all the time in Mombasa too, not just fairy tales but also poetry. It was he who had introduced her to Wordsworth, reading 'We Are Seven' to Helen at bedtime in a voice loud enough for Bella to hear.

'Can I help?' the woman behind the counter asks.

'I was looking for a copy of *Lyrical Ballads*,' Bella says, mentioning the first thing that pops into her head. She'd discussed her love of Wordsworth during her entrance interview at Oxford and his poetry was a favourite throughout her time at college. It made her feel closer to her dad.

The woman finds a copy and hands it to her, standing back, clearly concerned. Bella must look a fright. She takes the book and flicks through it, stopping at various poems, including 'The Mad Mother'.

A fire was once within my brain;
And in my head, a dull, dull pain;

It's no use. She can't focus on anything. Putting the book down, she walks out of the shop and heads up towards the train station, hugging the brick wall of a pub, ready to turn away if the police car appears again.

The station is still deserted when she reaches it. She walks onto the platform and sits back down on the bench. If the police appear in the car park opposite, she can slip away down the path. Checking that she is still alone, she reassembles her phone and is about to look for messages from her mum when it rings. She almost drops the handset in fright. No caller ID. Is it her mum? Her number always shows up but she might be calling from someone else's phone. Or is it the police? She lets it ring out. The phone pings with a message and she accesses her voicemail:

'This is DC Strover from Swindon CID. I interviewed you last night at the Slaughtered Lamb pub. Could you give us a call back – something's come up that you might be able to help us with.'

Bella listens, heart thumping, as Strover gives her mobile number. *Something's come up.* Her call to Crimestoppers? Shit, she told Strover about Erin, just as the detective was leaving, how her mum had once beaten up her dad. A siren goes off, making her jump. It's the level-crossing barriers, starting their shaky descent, but the sound takes her back to her first year at college, waking up in the middle of the night to the wail of a fire alarm.

'Everyone out of their beds!' a voice of authority had announced over the loudspeakers. 'Everyone out of their beds and assemble downstairs immediately, outside the main entrance.'

The first time it happened, she slept through the alarm and had to be woken by a porter – unfortunately, the same one they'd shut into Erin's room a few weeks earlier.

'Get your fat arse downstairs,' he'd shouted, on a final safety check of the college building. They could be cruel when no one was looking, not the quaint, bowler-hatted gents of Oxbridge folklore.

Bella looks around the station. She needs to calm down, compose herself. Recapture the mindset that got her through her final examinations. Her train is about to appear. There's no need to panic. But then, as the level-crossing barriers shudder into position, the police car appears on the opposite side of the tracks, slipping into a space in the car park. Have they seen her?

She is about to run back down to the high street when the train noses around the corner, hiding her from the police car as it draws into the station. The doors take too long to open. *Come on.* And then they are sliding apart and Bella steps on board. What if the police have asked the driver to wait? The blood is so loud in her head, beating on her eardrums like an enemy at the gate. *Close the fucking train doors.*

Two police officers appear on the bridge. Bella slinks down in her seat, hoping she's got this wrong, that they're just walking into town, but the officers step onto the platform and move towards her, checking the train windows as they go. Bella looks around, sees the loo at the far end of the carriage and makes a dash for it. A moment later, she closes the loo door, checking first that the police haven't seen her. They are at the seat where she was sitting a moment ago, shielding their eyes from the sun as they press their faces against the greasy window.

Remember, the police are not your friend.

47

Silas

'Any luck?' Silas asks, as Strover comes off the phone. They're sitting in the major incident mobile command vehicle, parked up outside the front of the hospital.

'Straight to voicemail,' Strover says. She is keen to talk to Bella, convinced that she had a friend called Erin – the name of the second victim left by an anonymous caller to Crimestoppers. 'I've also rung a mate at Hackney police station, someone I trained with. Owes me a favour – he's going to pay her a visit.'

Silas turns away, unconvinced. A uniform comes into the command vehicle as two others step outside. He's got more important things on his mind, like overseeing the ongoing lockdown of a large regional hospital. And everyone is on the back foot, playing catch-up. The ICU has become a major crime scene, but there's no sign of the fake doctor and, as yet, no security photos of him. Dr Armitage is also not answering his phone. His wife last saw him this morning at their home in Ogbourne Maizey, when he left for the hospital. At least Noah is still alive.

'It's not the biggest lead we've ever had,' Silas continues, looking again at the message about Erin. Crimestoppers does invaluable work – it's led to the arrest of more than 145,000 suspects – but sometimes he wishes it wasn't quite so strict about anonymity. If they could just provide him with the caller's phone number, for example... But he knows it's an impossible request, the reason the charity's so successful. They never disclose details of callers.

'Bella definitely mentioned that she had a friend called Erin,' Strover says.

'So does my wife,' Silas snaps, suddenly irritated. 'We're clutching at straws.'

'Maybe that's why Bella was in Wiltshire,' Strover continues, undeterred. 'Looking for her friend.' Silas remains silent.

'Just a hunch,' Strover adds, using one of Silas's favourite expressions. She's clever like that, plays things back at people. Subliminal flattery.

'Try speaking to her, by all means, but it's not a priority,' Silas says. He doesn't want to discourage Strover from having hunches, which have served him well over the years, but the death of the third victim, the 'zombie' patient, is their main priority. Silas is now SIO of a triple murder inquiry. And, as the media have gleefully pointed out, three deaths technically classifies the murderer as a serial killer. Finding the Range Rover driver and passenger who were brought in to hospital will help. The farmer involved in the traffic accident has given a statement, which should be with Silas any minute. Forensics is also combing the damaged Range Rover for evidence.

'I contacted Bella's newspaper,' Strover continues. It's

hard not to admire her doggedness. 'She isn't on the staff, but someone in the post room recognised the name, said she was a junior work experience.'

'Not exactly a top investigative hack then,' Silas says. 'Why would they send someone on work experience to get Jim to talk about Porton Down? She's irrelevant, met Jim at the pub by chance.'

Before Strover can reply, Silas's phone rings. It's Malcolm.

'Sorry to hear about your "zombie" victim,' he says, still clearly disapproving of the moniker. 'All sounds a bit careless.'

Silas rolls his eyes. The last thing he needs is a lecture from Malcolm.

'Now that he's dead, he's my problem, of course, not the consultant's, who I've just been speaking to,' Malcolm continues. 'The initial toxicology report found enough tetrodotoxin in the victim's system to slay a shire horse.'

'So your theory might be right, then?' Silas asks. 'About someone sending a message.'

'All I'll say at this stage is that whoever poisoned him clearly wanted the tetrodotoxin to be found by pathology,' Malcolm says, pausing. 'And yes, tetrodotoxin will always be associated with zombieism in Haiti, despite considerable medical evidence to the contrary.'

'Could it have caused his death?' Silas asks. There's a chance the fake doctor was just checking on him in ICU, to make sure he was dying.

'I doubt it,' Malcolm says. 'His condition was apparently stabilising. Tetrodotoxin attacks the body by blocking sodium ion channels. His was starting to grow new ones.'

'You mean he could have made a full recovery?' Silas

asks. The thought that he might have been able to question the victim is too frustrating to consider.

'Quite possibly. It's a bit like recovering from classical nerve agent poisoning. It can happen. Witness the Skripals, Alexei Navalny.'

Silas's ears prick up at the mention of nerve agents. Jim's analysis of the crop circles is still haunting him, the mention of BZ, or buzz.

'Have you ever had any dealings with Porton Down?' he asks.

'Not that I can tell you about,' Malcolm says. 'I get the odd call, most recently asking me to analyse the ballistic injury patterns after a pig's head had been shot at through a military helicopter windscreen. Why?'

Silas wished he'd never asked.

'I'm just wondering if there could be a link between these coded crop circles and various chemical weapons that have been tested at Porton.'

'Interesting. And frankly disappointing, Silas. I thought a respectable detective like you would be above such nonsense.'

'What nonsense?' Silas asks defensively.

'Conspiracy theories about Porton Down. Sensationalist paranoia.' A short, derisive laugh. 'Did you know there's an official statement on the government website confirming that no aliens, either alive or dead, have ever been taken to Porton Down?'

'Did I mention aliens?' Silas says. He doesn't want to seem a killjoy but he's in no mood for banter right now. 'We just have to rule certain things out, that's all.'

'Actually, I gather there is a bit of unrest over at Porton.

The novichok case could have been handled better. Politics and science have never been easy bedfellows. But leaving dead bodies in crop circles? That's not really the British way. A bit *de trop*. More Moscow's style, I'd have thought.'

'I'll call you later.'

Silas ends the conversation with Malcolm just as Strover comes off the phone.

'Give me some good news,' he says. All this talk of Porton Down, novichok and Moscow is giving him bad flashbacks.

'Just spoken to the officer who's sending over the RTC report,' Strover says. The glint in Strover's eyes is promising. 'Seems like the farmer's on the ball. He took a photo on his phone of the other car involved – the one he says caused the accident and drove away. The officer's already checked the number plate and it belongs to Jim Matthews.'

Silas sits back. Jim's less of a fantasist than he thought. Except that last night he was claiming the Range Rover had tried to make *him* crash. Now he's the one accused of causing an accident. 'And the farmer thinks that Jim was actually responsible for the RTC?' he asks.

'He was certain. And ANPR's already come up with a result – Jim was picked up heading south out of Salisbury a couple of hours ago. An obs request has been put out across all southern counties.'

48

Jim

Jim doesn't head straight back to his dad's old home in Swanage. He needs to confirm that the Range Rover he spotted on the seafront is here for him. It seemed identical to the one he saw at Hackpen Hill, which was similar to the vehicle involved in the accident. If that's the case, there must be at least two Range Rovers, as one of them would have been badly damaged by the tractor. He also needs to pick up his bag from the boot of his car. Checking the busy car park, he removes it and walks up towards the Catholic church and the road where he used to live.

He's still five hundred yards away, cresting a small incline, when he spots the Range Rover up ahead, beyond the house and driving slowly towards him. It's the same car he saw at Hackpen Hill, he's sure of it. Trying not to attract any attention, he slips into the small park that sits between the road and the beach below, and waits behind a thick hedge, winking at a young boy nearby who eyes him with suspicion.

'Shhhh,' Jim says, finger to his lips. 'Hide-and-seek.'

Jim used to be so good at hide-and-seek at school that he was never found. Or perhaps no one was looking. The boy nods solemnly as the Range Rover drives past. Jim doesn't get a clear look at the driver but he recognises the number plate. It's definitely the same vehicle. Which means they must have turned around in Marlborough and followed him down to Swanage. But do they know about his old house or are they just driving through the town looking for him? It's a busy road, a popular cut through.

He waits in the park for a couple of minutes before making his way to the house. No one is around. The neighbour on one side is elderly, the other is often away travelling. Jim's stayed here occasionally in the past twelve months, when he's come to visit his dad for a weekend. He walks around the back and across the parched lawn to the once brightly coloured summer house, pressing his nose against the glass. Plants are climbing up through the floor and walls and a solitary blue plastic boules ball lies cracked in the far corner. He used to spend hours in here with his fossils. They seemed to offer a break from thinking, a way to calm his brain.

He walks back across to the house, takes a deep breath and unlocks the side door. A wave of sadness hits him as he enters the kitchen. The place is empty but there are memories everywhere. Dad's study, straight ahead, a mark on the wall where the weather station was once installed. The sitting room to his left, where he learnt to play the piano. The old metronome imposing order on his music – on his life. And the nanny's room to the right, a constant reminder that his mum had died too young.

Five minutes later, he's upstairs at his old desk, sitting in the window that looks out over Swanage Bay. Everything's been erased from here, all traces of his childhood apart from a tiny piece of graffiti that the decorators missed – 'Phuck Physics', carved into the edge of the windowsill. It had felt so rebellious at the time. He'd always preferred chemistry, much to his his dad's irritation. In truth, he found all the sciences easy at school, taking the exams a year early. Advanced maths too.

He opens his laptop and scrolls through a file called 'Modern Maddison' – the same file that he downloaded onto the USB and gave to Bella. Has he got enough proof yet? Was it a mistake to give the USB to her? She won't be able to open it without the password. It's on a draft text, ready to send to her in an emergency.

He gets up from the desk and looks out of the window. The Needles on the Isle of Wight are bathed in warm afternoon sun. He always preferred Swanage after the summer, when the place was reclaimed by the locals and beat to a calmer rhythm. No more queuing for cod and chips at the Fish Plaice on a Friday. If only he knew who was sending him messages through the crop circles. How much do they know about his own research? If the first circle is the incapacitating agent BZ, the second one LSD, and the third one the nerve agent VX, they are both singing from the same song sheet. All three substances have played a significant role in Porton Down's controversial history of human trials.

Jim leans back, mulling over the implications. It's not just in the past – old habits die hard. Twenty thousand human guinea pigs have passed through Porton Down since the place was founded in 1916 and its volunteer programme is very much ongoing. Young squaddies continue to take part in tests to improve protective clothing and medical countermeasures for today's armed forces, most recently for women on the battlefield. A couple of days away from their regiment, a chance to earn some beer money. What's not to like? But there's another story to be told, one that Jim discovered during his three years at Harwell, the affiliated site where no one would ever think to look. It's here that the unethical experiments continue as if the death of Ronald Maddison in 1953 had never happened. BZ, LSD, VX and more.

Jim will let Bella tell the world but first he needs to discover who is sharing his concerns so publicly. Closing his laptop, he's about to go downstairs when the front doorbell rings. Nobody calls here any more. All mail is redirected and it's known locally that the house is empty. Walking quietly across the landing, he looks down onto the street. A black Range Rover is parked up outside the house.

49

Bella

Bella checks up and down the street and opens the front door of her house in Homerton. It's been a stressful journey back to east London and she's calmed by the familiar smell of incense. She stayed in the train loo until Reading, only coming out when a desperate passenger knocked on the door and pleaded with her. There were no police on the train or at Paddington, but a patrol car drove past as she walked from Hackney Central station.

'Mum, I'm back,' she calls out in the hall, but without much confidence. She has put the battery in her phone half a dozen times and tried to ring her.

The house feels empty, uneasy. Interrupted. Her mum's reading glasses are on the kitchen table, next to a copy of the book on Somalia that she was engrossed in two nights ago. Beside it, dried orange peel. She once showed Bella how to light a fire with it. A newspaper is folded, unread. She walks over to the sideboard. An unused herbal teabag in a mug. The kettle warm. She must have left in a hurry.

A man's voice, faint but close. She stands still, straining

to hear, pulse picking up. Is it coming from another room? She sees the radio on the dresser and walks over. It's on, very quietly. The World Service. Her mum either has the radio on loud or off. It's not like her to leave it like this. Did she turn the radio down when she was trying to hear something?

Bella walks back through to the hall, scalp tingling. The door to where the two lodgers are staying is closed. She knocks once.

'Hello?' she calls out. 'Anybody home?'

Silence. She tries the door but it's locked. Her mum insisted on them having some privacy when they moved in. They're usually out during the day, either helping at the migrant centre or at Ridley Road market in Dalston. Slowly she climbs the stairs, glancing at the African-print drapes hanging over the banister.

'Mum?' she says again, less confident now.

Her bedroom's empty. Bella checks the spare room – Helen's room. Her mum left it unchanged for months after she emigrated, but her stuff is packed away in the loft now and the room is used for guests. At least, it was when Bella was living at home before college. Her mum's former aid agency colleagues passing through London, old friends from Kenya. Should she ring Helen now? It's the middle of the night but she'd understand.

She walks down the landing and looks in on her own room. Nothing's changed – except for a typewriter on her desk. And a handwritten note beside it.

Been rummaging in the attic and found this. Thought you might like it, now you're an intrepid reporter. It's Dad's trusty old Remington, the one he wrote all his

stories on. He would have loved the thought of you
using it to file your first piece for a national newspaper.
So proud of you. Mum xx

At the end of the note, she's drawn a flower. Bella presses the keys of the typewriter, watching the levers swing forward like miniature golf clubs. Her dad's staccato typing used to echo through their house from his study in Mombasa. She and Helen knew to be quiet when the typing sped up. It meant a deadline was looming. He could have switched to using a laptop but he preferred his typewriter. Steeped in old-school journalism, he insisted on reading out his story to copy-takers in London and Washington, using cheap burner phones bought in the market that were untraceable. It was safer than using hard drives that could be hacked, emails that could be intercepted. He had made too many government enemies, exposed too many corporate crooks.

What would he do in her position now? He wouldn't have taken no for an answer about Erin's illness. And he would definitely have investigated her death. Whatever the personal risk. Once he was away for two whole months. A friend of his, a human rights campaigner, had been murdered in south Mogadishu and he wanted to track down his killers, expose them. Bella has since discovered that when the investigative article finally came out, it won awards and he was hailed a hero. Except by the friends of the arrested killers, who got their revenge in the end.

She's about to return downstairs when a thought slips into her mind and the room seems to darken, as if the day has momentarily lost power. Taking a piece of A4 paper out of her drawer, she inserts it into the typewriter, scrolling the page into position. Another memory: the way he let her open a new ream of paper, the fresh smell of vanilla. She starts to type, slowly, with two accusatory fingers.

```
PS Make sure you 'overhear' the man on
his own in the corner.
```

She pulls out the original letter from her bag and holds it next to the typewriter, her hand shaking. The font is identical.

50

Silas

'I've been doing some research into *folie à deux*,' Strover says to Silas, as they wait for the secure Zoom meeting to start. They are still in the major incident mobile command vehicle outside the hospital, which remains in lockdown. Silas can hear the police helicopter overhead, part of the net they've thrown across the site in an attempt to find the fake doctor.

'Anything interesting?' Silas asks, half turned away from his computer's camera, as if he's shy. The Zoom meeting is not great timing. He also doesn't like to see himself on screen. A reminder of how much weight he still needs to lose. Strover's suggested that he turns off his camera and mic and renames himself as 'Reconnecting' when Zoom meetings go on too long and he wants to leave.

'Literally, *folie à deux* means a shared psychotic disorder or delusion,' she says. '"Madness for two", or "double insanity". Although you can get *folie à trois*. Sometimes more. There was a case a few years ago in India where eleven members of the same family took their own lives. Delhi police suspected it was a case of *folie à famille*.'

'And how does that relate to our crop circle victims?' Silas asks, still no wiser why the zombie patient had whispered '*folie à deux*' in the night.

Before Strover can answer, two academics appear on his screen, popping up one after the other in quick succession. The female professor of mathematics from Cambridge, and a male chemistry professor from Imperial College, London.

Silas switches to gallery view and for a second, as he looks at the grid of images, he thinks he's starring in an episode of *University Challenge*. Except that he wouldn't be on the programme, having never been to university. After some cursory pleasantries, Silas cuts to the chase and asks for an update – 'in simple, layman's terms, please.'

'We have two things going on here,' the professor of chemistry says, sharing an image of the first crop circle, at Hackpen Hill, on the screen. The professor is balding, in his late forties at a guess, with a wry, world-weary tone of voice.

'Geometric patterns on the right and a spiralling binary sequence to the left. I'll concern myself with the former and allow my colleague to expand on the binary. From what we can tell, the two hexagons and surrounding lines would appear to represent the structural formula of a chemical compound,' the professor continues. 'But we're not sure at this stage about the exact molecular geometry. Parts of it appear to be incomplete – linear rather than 3D. Whoever made it might have run out of time.'

The connection isn't great so Silas tilts his laptop, hoping to improve the poor picture quality. Zoom is one of the many legacies of Covid-19. The force's bean counters now refuse to sign off travel expenses without first satisfying themselves that the visit could not have been conducted online.

'It's not clear from the pattern, for example, what exactly these represent' – he highlights the two hexagons – 'but the parallel lines on two of their sides suggest the telltale alternating double bonds of a benzene ring, which has six carbon atoms, each one bonded to a hydrogen atom.'

Silas feels the same pang of restlessness he used to get in chemistry O-level classes. He glances at Strover, who remains firmly focused on her screen.

'Which doesn't exactly narrow it down much,' he continues. 'There is a large class of aromatic compounds that contain benzene rings.'

'How about BZ?' Silas asks, remembering what Jim had said. 'The incapacitating agent?'

Strover calls up the details on her screen and shares the molecular structure. Silas had warned her that he was going to throw BZ into the mix, and had asked her to prepare a screenshot.

'Your colleague mentioned BZ earlier,' the professor says. 'It's a troubling suggestion – I won't ask why it's become a candidate because I suspect you're not allowed to tell me. 3-Quinuclidinyl benzilate does indeed have two benzene rings, but it's not clear, at this stage, that the rest of its molecular structure is depicted here.'

'So how do we set about confirming the compound in these crop circles?' Silas asks, keen to target the problem in hand.

'If I may come in at this point,' the professor of mathematics says, leaning forward. She is younger than the man, with short-cropped hair, and seems more enthusiastic about her subject and life in general.

'Be my guest,' Silas says, sitting up.

'The spiral pattern that adjoins all three crop circles is, I think, crucial,' she begins. 'Certainly more detailed. It appears to contain a binary sequence, moving out from the centre. We've seen this sort of thing before, most notably outside Winchester at a place called Crabwood Farm in August 2002. And in May 2010, at Wilton, near Marlborough, where Euler's Identity was encoded in a binary wheel.'

Silas looks up. To his surprise, he feels reassured by the mention of Euler's Identity. And thanks to Strover, he now knows that binary's ones and zeros can be converted into everyday characters using ASCII code. She's done her best to educate him, but he still feels out of his depth.

A pause while the professor shares a photo of a large crop circle and an even bigger face of a Hollywood-style alien, carved out of the wheat. Silas glances at Strover. Here we go. ET time. The alien is all big eyes and narrow chin, bearing an uncanny likeness to the figure in Munch's *The Scream*.

51

Jim

Jim stands on the landing of his dad's old home in Swanage, breathing hard. The back door is locked and there are no open windows. The house is secure, no need to panic. But how did the people in the Range Rover, parked outside, know that he's here in this deserted house?

Jim flinches as a bell rings and someone raps on the front door. A moment later, the letter box flaps open.

'We just want to talk with you, Jim,' a man says. It's a different voice from the person who rang him in the car. 'Have a chat about things.'

Jim tenses again, tells himself to relax. Did they go to his dad's apartment? If they touched him, harmed him in any way, he'll kill them. He needs to calm down. Maybe they're bluffing, don't know for sure that he's holed up in here. He's certain nobody saw him enter and he was careful not to turn on any lights.

'We need to know how you met Bella, the journalist in the pub,' the man continues, his voice echoing up from the empty hall below.

Unless they chuck a brick through the window, there's no way for them to enter. His dad was a stickler for security. And Swanage is too busy at this time of year for a break-in to go unnoticed. They could come clean that they are from MI5 and force entry. But that's not their style. And they won't have a warrant, not for this house.

'Did she arrange the meeting?' the voice asks. 'Or was it by chance? It's really important you tell us.'

Jim closes his eyes, shaking his head. They're on to him as a whistle-blower. It's the only explanation, the reason for such a specific line of questioning.

'We also need you to tell us how much you know about these crop circles. You must share it with us, Jim – any information you have, however small. Anything at all.'

A sudden thought occurs to Jim. Are these people aware that there are two whistle-blowers at Porton Down? Or do they think there's only one – and that he's responsible for the bodies in the circles? In which case, they'll try to charge him with murder as well as contravening the Official Secrets Act. Not that he's told anyone yet what he's discovered. Either way, he needs to keep them at arm's length.

He glances around the landing. The windowsill, now bare, was once piled high with every kind of Rubik's cube he could find. His record was twenty-five seconds, a good minute faster than his dad. He looks up and sees him coming out of his bedroom, red-and-blue-striped flannel dressing gown wrapped tightly around him, humming a tune as he completes the cube and looks at his watch. And then he's gone.

They shared a lot of good times together: restoring an old Mirror dinghy in a miasma of fibreglass fumes in the garage;

making a fuel-powered rocket that nearly reached the Isle of Wight. But it was a mutual interest in prime hunting that Jim misses the most, using free, open-source software. When the largest known prime number was discovered a few years back, by an IT professional in Florida, Jim rang his dad to tell him the exciting news.

'It's another Mersenne!' Jim had enthused, referring to a rare class of prime number.

'I can't find my cup of tea,' he had replied. It was then that Jim began to suspect something was wrong.

'We know you're in there,' the voice downstairs says again, bringing Jim back to the present.

No they don't, not for certain. They're guessing, taking a punt. But Jim doesn't feel any relief. Only anger. Last time Jim was paid a visit by these people, his house was ransacked and he was beaten up, knocked unconscious. That's the way MI5 does things. It likes to cut up rough, particularly with potential traitors. He feels a surge of rage rise through his body. Anger at his dad's cruel disease, his mum's early death, what they continue to do to people in the name of science at Harwell. How dare they behave like that? And what a nerve to turn up here, at his old family home, trampling back through time on his precious memories?

He moves to the wall, shaking now, and thumps his forehead against the hard plaster. The pain is fierce and immediate and he screws up his eyes, trying not to shout out. And then he does it again. And again. The discomfort will pass. It always does. Blood, warm and metallic, begins to seep down his face from the reopened wound. Enough now. He stands and listens, shaking, waiting for them to bark out another question. Silence. They won't stop him from revealing the truth. He leans towards the window again, careful not to be seen.

The Range Rover has gone.

52

Bella

Bella returns downstairs from her bedroom, still reeling from the words she typed out on her dad's old Remington. It doesn't mean that the anonymous letter was written on it. Or that her mum sent it to her. All old typewriters have a similar font, don't they? But it's left her even more anxious that her mum isn't here.

Why did she leave in such a hurry? As Bella reaches the hall, she spots another note, a scrap of torn paper, tucked under a wooden carving from Kenya. She must have missed it earlier. Bella steps forward and picks it up.

Have to go out. Stay safe, Mumxx

It's not written with the same care as the note upstairs in her bedroom. Someone else could have almost written it, but she can still see traces of her mum's floral style in the dashed-off words.

She walks over to the landline phone in the kitchen and rings her number, holding the note in her trembling hand. The call goes straight to voicemail and she leaves a message.

'Hi, Mum, where are you? Ring me when you get this. Got your note. Hope everything's OK.'

Should she call the police? It's so unlike her. She never writes notes like that. It also looks as if it's been hidden behind the carving, out of sight. Her notes are always thoughtful, considered, like the one upstairs, signed off with a small drawing, usually a flower, sometimes a bird. But then she checks herself, remembering Jim's words of warning about the police.

She walks into the sitting room and looks out of the window. The car is parked outside, where her mum always leaves it. Back in the kitchen, she picks up the landline handset again and scrolls through the list of recent missed calls. Several are from Gladys, a friend of mother's who works at the migrant centre. And then she sees Dr Haslam's name. It's the last incoming call before the missed calls. A short, twenty-second conversation. Of course. They're together, that's what's happened. Her mum can be impulsive like that. He rang, asking if she wanted to meet for a quick drink, and she dropped everything to join him. She wouldn't have explained in her note who she was meeting, knowing how disapproving Bella's become of Dr Haslam.

Bella pulls out her own mobile and slots in the battery. Lots of missed calls from Jim. She was going to text him earlier on the train, to say that Rocky was fine and that she'd left the house, but she didn't want to risk turning on the phone. She dials his number and waits.

'You OK?' he asks. 'I've been trying to call you.'

Bella closes her eyes with relief. She didn't realise how

alone she was feeling, how comforting Jim's voice would sound. For a few minutes, she'd forgotten about Erin but now her death comes crashing back into her consciousness.

'I'm fine,' Bella says, looking around the abandoned kitchen, tears in her eyes. Why's she lying to him? She's not fine at all. She should just come clean, tell him about Erin.

'You got my message about Rocky?' he asks. 'Left the house OK?'

Jim doesn't sound himself. Or maybe it's her. She's struggling to concentrate.

'Yeah. I was going to reply,' she says, wiping her eyes, trying to keep it together.

'It's cool. Where are you?'

'I'm in London. Where are you?' She can hear seagulls. She used to love the sound of the sea.

'Swanage. I've just been seeing my dad.'

'Was he OK?' she asks. A pang of recognition. She's sure Swanage isn't far from Studland. Distant memories of arriving there on a steam train with her dad. He loved old trains, even proposed to her mum on the overnight express from Mombasa to Nairobi.

'Not really,' Jim says. 'He's got dementia.'

'I'm sorry, I didn't know.'

Bella hears her dad's typewriter echoing through the house. Tries to imagine what he'd do now. He would focus on the story. The heart of things. If she hadn't received the anonymous letter, she wouldn't have gone to Wiltshire, met Jim, heard so much about the crop circles. They're the key to explaining Erin's death. The talk is of them being connected to Porton Down, where Jim works. It's all too much of a coincidence to ignore.

'You need to tell me about Porton Down, Jim,' she says, trying to sound confident.

'You got the USB?' he asks.

'I got it,' she says, remembering how she fished it out of Rocky's vivarium. 'But I can't open it. Tell me what's going on. Please?'

'I can't – not yet.'

Bella starts to well up with anger. Jim doesn't know about her connection with Erin, that her interest in the crop circles is no longer just professional.

'Why not?' she asks, trying to control her voice. 'It's all linked back to these crop circles, isn't it?'

'Bella, it's really not safe to speak on the phone,' he says.

'Jim, she's dead,' Bella interrupts, almost shouting now. 'Erin, my best fucking friend at college. It was her body that was found in the second crop circle. Her eyes that were pecked out by birds.'

53

Silas

'This is Crabwood, 2002,' the professor of mathematics says on Zoom. Silas looks again at the image of the Hollywood extraterrestrial and shakes his head, glancing at Strover beside him.

'Ignoring the ridiculous alien,' the professor continues, 'which isn't hard to do, the spiralling circle really is quite intriguing. If we follow the binary sequence out from the centre in an anticlockwise direction, we see that it is broken down into a series of blocks, each one with eight pixels in it. These pixels are a patch of either flattened crop or standing crop. And each block of eight pixels is separated from the next block by a divider – a smaller patch of standing crop. We can take this sequence to correspond to the eight-bit ASCII code used by computer programmers the world over. Flattened crop equates to zero, standing crop equates to one. Is everyone still with me?'

Silas nods at the screen, trying to keep up with what she's just said. He hasn't concentrated so hard since he was in shorts.

'The groups of eight binary digits can then be converted into ordinary text using ASCII?' Strover asks. 'Like the Euler's Identity circle at Wilton in 2010?'

If they were playing *University Challenge*, Silas would be glad to have Strover on his team. In fact, she could be captain.

'Exactly,' the professor says. 'First up, we need to take one of the eight-bit binary numbers – 0100 0010, say – and convert it into its decimal equivalent, 66. We then see what it corresponds to in ASCII's character set. In this case, the capital letter "B". The message in the 2002 circle was relatively straightforward to decode – and really not worth the effort – but unfortunately it's not so simple with these messages. We've run the binary codes through an ASCII converter and—'

'And what?' Silas interrupts. 'What does it spell out?' He's impatient for results today, nervous in the company of academics.

'Nothing, I'm afraid,' she says. 'We get a string of meaningless text.'

Silas sighs, unable to disguise his disappointment.

'Does the text need to be decoded too?' Strover asks. 'Like a double lock?'

The professor nods. 'That's where we're at. I won't bore you with the details, but we think that whoever is behind this has used a Vigenère cipher – a traditional way of encrypting text by using a series of Caesar ciphers that are based on a key word.'

'Polyalphabetic substitution?' Strover asks.

Silas gives her another look. Five bonus points to Swindon.

'Precisely.'

The mathematician doesn't bat an eyelid at Strover's knowledge. Silas knows he shouldn't either. It's what comes from a university education.

'The basic idea behind the cipher is to deny codebreakers the opportunity to do frequency analysis on certain letters such as "e", which is the most commonly occurring letter in English,' she continues. 'The weakness is the key word. We don't need to know exactly what it is to unlock the code but if we know its length, we can break the ciphers individually by calculating how many different alphabets have been used.'

'And that's what you're working on?' Silas asks. Despite her lack of instant answers, Silas likes the professor, the way she doesn't talk down to him and Strover.

'Correct,' she says. 'We're currently using something called the Index of Coincidence to determine the key word length. We've also been in touch with some American colleagues at the National Security Agency in Maryland – the Vigenère cipher was once used to encrypt part of a coded message on a sculpture at the CIA headquarters in Langley, Virginia. Its unveiling was a challenge to cryptanalysts everywhere – and the guys at NSA beat everyone to it back in the 1990s.'

'Keep an open mind about BZ,' Silas says. 'Maybe the letters will turn out to be its formula.'

'Absolutely,' the mathematician says. 'They could well be numbers or letters or a mix of the two.'

'And any idea how long it might take to decrypt this Vigenère cipher?' Silas asks.

'The guys at NSA used a pen and paper and took

THE MAN ON HACKPEN HILL

forty-eight hours,' she says. 'We've got supercomputers these days so it shouldn't be long. Having said that, there's one part of the CIA sculpture code that still hasn't been deciphered by anyone – more than forty years after it was made.'

54

Bella

For a moment, Bella wonders if Jim has hung up. Or has she just shocked him into silence with her revelation about Erin?

'Jim?' Bella asks, glancing around the kitchen. She mustn't talk for long on her mobile. It's too risky.

'I'm here,' he says, his voice barely audible. 'How do you know this?'

'I know because she'd covered her arms in feather tattoos – like the body they found. She had a rook behind her ear too. I thought she was recovering in hospital but...'

She can't go on any longer and starts to sob again.

'I'm so sorry,' Jim says, struggling to find the right words. 'I had no idea. I wish I... I wish I was there with you.'

'You really need to tell me what's going on here,' Bella pleads. She's played her trump card, told him about Erin. There's nothing more she can say to persuade him to reveal all.

'I will, I promise.' He pauses again. 'Did your friend...

did Erin ever act strangely? Out of character? Like she was tripping or something?'

Jim's voice has changed, become more matter-of-fact. Analytical.

'Why?' Bella asks, wondering how he knows, dreading what he might say next. Erin was queen of strange at college, always tripping out.

'These crop circles, they're complex, coded patterns,' he says. 'No one seems to know what they are. The hexagons hint at molecular structures and the wheels, the binary sequences... I'm sure they contain their chemical formulas. I've only seen pictures of the second circle briefly, the one Erin was found in, but I think I know what it means.'

'You do?' Bella asks. Why hasn't he told her this before?

'It looks like a representation of lysergic acid diethylamide. LSD. Acid. I think the first circle represents BZ, a psychochemical weapon developed by the Americans in the 1960s and last used by Assad in Syria. The third one I'm sure is VX, the most potent nerve agent ever made. All three are kept at Porton Down.'

This time it's Bella's turn to remain quiet, thinking of Erin's body lying in the flattened wheat. Erin dropped a lot of acid at college.

'You still there?' Jim asks.

'Erin had issues,' Bella says quietly. 'Took loads of drugs. But she's never been to Porton Down, never had any dealings with a place like that. She grew up in Dublin and London and then spent three years at Oxford on a full scholarship.'

'I'm not saying she's been to Porton,' Jim says. 'It doesn't always work like that.'

'What doesn't?' Bella asks, unable to contain her frustration.

'Have you rung the police?' Jim asks, ignoring her question. 'About Erin?'

'Yes, I mean no,' she says. She won't be able to cope if he's cross with her. 'I rang Crimestoppers but they don't—'

'Did you leave your name? Tell them Erin's? It's important you tell me, Bella.'

He sounds more concerned than angry. 'Not my name, no,' Bella says, again regretting that she'd made the call to Crimestoppers. 'I just said that my friend was called Erin.'

She can hear Jim sigh. 'That might have been a mistake,' he says. 'A big mistake. You've got to be very careful with these people. Believe me, Bella. I know what I'm talking about. Take the battery out of your phone as soon as we hang up.'

'I've been doing that,' she says, looking down at her mum's scribbled note on the hall table.

'Can you get down to Swanage?' Jim asks.

'Me? Swanage?'

She's only just returned from Wiltshire, but she has no desire to remain in this empty house. She glances at the hall table again. At the car keys. If her mum can drop everything for a romantic tryst with Dr Haslam, she can go to Swanage. Her mum never uses the car in London. Bella would prefer to take the train but it will take too long. Her only other worry is Jim. Can she trust him? He seemed genuinely shocked by Erin's death but how much does he really know about the crop circles? What if he was involved in their making in some way?

'I might not have much longer,' Jim says. 'They're here, in

the town, but they won't find me. I know this place better than them, grew up here. Bring your laptop – and the USB. And turn off your satnav. Don't use anything that could be hacked to track you. Call me when you get near and I'll tell you where to meet.'

55

Silas

'Play it again,' Silas says, eyes fixed on the TV screen. He's back in the estate manager's office at the Great Western, watching CCTV footage from the front of the hospital. The manager called him fifteen minutes ago, asking him to come over as quickly as he could. Silas obliged, happy to avoid another Zoom call. The site is still in full lockdown. Uniforms everywhere, patrol cars lined up at the front of the hospital, police helicopter overhead.

The manager leans forward to play the footage again. It's taken from earlier in the day, when the two men involved in the Range Rover accident discharged themselves from the hospital. One of them has his arm in a sling, the other is wearing a neck brace. Silas watches as they walk out of the main entrance and stand in the pick-up area. Ten seconds later, after a lot of glancing around, they board a bus that's swooped up through the car park.

'What else?' Silas asks, turning back to the estate manager. It's good, but Silas senses there's more.

'We've only just seen this, but I'm sure it's him,' he says.

Silas can detect the suppressed excitement in his voice, trying to remain professional.

Again, the footage is from earlier. Silas watches as a man exits the front entrance. It's the 'doctor' who passed them in ICU. He walks over to the same pick-up area, glancing at his watch. After two minutes, he pulls out his phone, looking around him again. He then makes a call, talking animatedly, before walking off camera.

'We pick him up again here on another camera, running around the back of the building,' the estate manager says.

The man is heading towards A & E and a queue of ambulances.

'Whoever was meant to collect him at the front didn't turn up,' Silas says, sitting back.

This is good, the sort of break they need.

'It gets better,' the estate manager says. 'We've just found this.'

Silas leans forward as the manager plays a third piece of footage.

'Where's that?' Silas asks, watching as the man uses his pass to enter through a door.

'It's the lower ground floor – staff entrance.'

'How long ago was this?'

'Ten minutes?'

'So he could still be on site?' Silas asks, mind racing.

'There are two visitor lifts to the lower ground floor, two service, one clinical – we've isolated all five. And I've got my security guards on every exit. He hasn't come out of any of them – we've checked the footage.'

'And what's down there?' Silas asks.

'The training academy, facilities management, plant rooms,' he says, pausing. 'And the mortuary.'

Silas looks up. He had been allowed as much time as he wanted to stay with his father after he had died. But as soon as Silas had left, his body was washed and taken straight down to the mortuary. Mel had told him later. 'Where is this entrance then?' he asks.

'Just beyond A & E.'

'Strover?' Silas calls out through the half-open door.

'Yes, sir?' she says, coming off the phone as she returns to the room.

'We need four uniforms *now*.'

56

Bella

Bella takes one last look around the house, in case she's forgotten anything, and heads for the front door with her bag. She's packed enough clothes to stay a few days in Swanage. Jim sounded stressed and she has no idea what he has in mind. All she knows is that she needs to get to the bottom of what happened to Erin and Jim is currently her only lead.

The landline rings just as she opens the front door. Bella puts down her bag and walks back into the kitchen. Should she answer it? What if it's Dr Haslam? She picks up. Gladys, from the migrant centre.

'Hello?' Bella says nervously.

'Is that Anne-Marie?'

'No, it's Bella.'

'Bella! You two sound so similar it's spooky.'

'Do you know where Mum is?' Bella asks, interrupting her.

'Trying to get hold of her myself,' Gladys says. 'We down two members of staff today and it's our big early dinner in an hour. Right now it's just me and the cat and she ain't so good at serving the soup.'

Gladys is one of the migrant centre's larger-than-life heroes, always upbeat and full of laughter. She's the de facto boss of the place, loud and maternal, and loved by migrants and co-workers alike.

'Do you know where she might be?' Bella repeats, failing to disguise the emotion in her voice. The fact that Gladys has been trying to get in touch with her mum too makes Bella doubt whether she's with Dr Haslam.

'What happened, girl?' Gladys asks.

'I don't know where she's gone,' Bella says.

'Listen, you know what Anne-Marie's like. She's probably on her way over now, met someone in the street and started to chat. Cha, always ah chat.'

Gladys is right. Her mum talks to anyone and everyone. Bella's just been overreacting, letting her imagination get carried away. She's having a quick drink with Dr Haslam before heading over to the migrant centre. But then she remembers the scribbled handwriting, the way the note was tucked away under the carving.

'You're probably right,' Bella says, unconvinced.

'I know I'm right. And if she's not with you at home, she's on her way. In the meantime, I was thinking... Might you be able to drop by and help out round here? In case she really has forgotten it's our big day?'

Bella glances at her watch. She worked in the kitchen when she first came back from uni, helped lay out the tables and chairs, enjoyed the camaraderie, but she should be heading to Swanage. Her best friend's dead and she needs to know why. On the other hand, if her mum's on her way over there, it would be good to see her, thank her for the typewriter, talk to her about Erin, ask about borrowing the car. And it might help to take Bella's mind off things, allow her to see the wood for the trees. Focus on others for once. Her mum was always saying how her time at Oxford was so self-centred. It will only be for an hour and she can drive on to Swanage from there.

'OK,' she says. 'I was just heading out in Mum's car. I'll be there as quick as I can.'

'You're a star, girl,' Gladys says. 'Park in the yard round the back.'

Twenty minutes later, Bella is ladling out portions of curried coconut chickpea soup to a string of newly arrived migrants from Belarus, Somalia and Eritrea. It's good to be back here again. She likes the work and is proud of her mum, how she's set up a much-needed local facility. At one point, she offers to hold a baby while the mother, from Ethiopia, talks to one of the professional advisers who have volunteered their services for the afternoon.

'You're a natural,' Gladys says, watching as Bella hitches the baby onto her own hip and carries on serving the soup. 'Don't know what we done without you – this place is bursting at the seams.'

Gladys looks down the long queue. Bella follows her gaze, smiling at the array of people of every colour and creed, the way Gladys relishes the 'r' in 'bursting'. They'll all get help today with immigration paperwork as well as their housing applications. Some of the advisers are waiting at Formica tables spread throughout the hall, others are mingling.

'Good to see you looking so well too,' Gladys adds, more quietly now. 'After all you've been through. I remember when you were this high' – she puts a hand out at knee level – 'and holding on to your mum's pretty skirts.'

Before Bella can ask Gladys what she means by 'all you've been through' – her dad's death? – she spots a young policeman entering the door at the far end, making his way politely but purposefully through the queue. Bella swallows hard. Gladys has spotted him too – and Bella's reaction. She doesn't miss much.

'Maybe you need to go get us some more soup,' Gladys says, glancing again at the officer and then at Bella.

'You think so?' Bella asks, peering into the big stainless steel pot in front of her. It's still half full.

'I think so,' Gladys says, giving Bella a wink. 'I'll handle things out here. Give me the baby.'

'Thanks,' Bella says, passing the baby over. The police officer is close now, making his apologies as he steers a course through the melee of people.

'And if anyone asks,' Gladys says, 'you're not here, right?'

'Never have been,' Bella says. 'Thanks.'

'Your secret's safe with me,' Gladys says. 'But you tell me everything later. You promise?'

'Promise.'

Bella turns to take the pot back to the kitchen, her heart racing. She's sure the policeman has come to ask about her. Unless he's here to break some bad news about her mum.

An old man she's met before, from Ethiopia, is wiping down the surfaces in the kitchen as she places the pan beside the stove. The man only speaks Amharic. She puts her finger to her lips, smiles at him and slips outside into the back yard, closing the door behind her. A moment later, she hears Gladys's voice in the kitchen.

'I'm in charge today and I know we don't have no Bella working round here,' she's saying, her voice loud and confident. 'I've never even heard of a girl called Bella. I wish I had, mind – we could do with a pair of extra hands right now.'

Bella stands and listens outside the door. Thank you, Gladys.

'I just need to check,' a man's voice says. The officer's, presumably.

'What is this, some kind of a fed raid?' Gladys asks, only half joking.

'Nothing like that,' the officer says. 'Nothing to raid, is there?' His tone has turned hostile. 'If she does show up, give me a call on this number.'

'Is she in trouble?' Gladys asks.

'I'm not sure why you'd care,' the policeman says. 'If you don't know anyone called Bella.'

Bella closes her eyes. Has Gladys blown it? She should go now, get in the car and drive to Swanage.

'Ignore me, I'm just the nosy type,' Gladys says, trying to recover the situation. 'Want the curried soup on your way out? You could do with some meat on them skinny bones.'

'On a diet,' he says. 'Smells good though. What's through there?'

'Nothing much.'

Bella can hardly breathe. Keep going, Gladys. Do something – anything – to distract him.

'It's where we put the trash out,' Gladys continues. Bella can hear the sudden tension in her voice. 'You don't want to go out there – it's well renk on a hot day like this.'

'I'll decide where I look – if that's OK with you.'

57

Silas

Silas stands outside the entrance to the lower ground floor of the Great Western, around the corner from A & E, with Strover and four uniforms. The only time he's been down here before was to have lunch with Mel at Bookends, the staff canteen in the training academy. She was tired in those days, angry with him and her job, Covid-19 and the government.

'I'll go first,' Silas says, swiping the entry card that the estate manager gave him earlier. The manager stands back as Silas walks in, followed by Strover and the four officers, leaving a security guard to keep watch.

'You take two officers and check the restaurant, I'll do the kitchens,' Silas says.

It doesn't take them long to establish that the fake doctor is not hiding in any of the obvious places. What bothers Silas is why he didn't leave the site when his lift failed to turn up. Maybe he felt too exposed – there are cameras everywhere in the sprawling car park and exit road that winds down through it. Is his plan to lie low until someone can collect him?

Silas checks the two plant rooms and then regroups with Strover outside the door to the mortuary. His dad was very peaceful in his last few days, ready to die, but it had still been difficult for Silas to let him go. There was so much left unsaid but Silas had checked himself, knowing that a heart-to-heart might have made him feel better but would only have exhausted his dad.

There's one person in the mortuary, a technician who has been waiting for them. It's not exactly a warm welcome. The man's unhappy about the disturbance, and the suggestion that someone might be hiding in the mortuary without his knowledge. He's also not big on eye contact as he stands in the doorway, blocking their way.

'Have you been in here all the time?' Silas asks.

'I went out earlier, for a cup of tea,' he says, nodding down the corridor in the direction of the staff canteen. 'Otherwise, yes. We're quiet today.'

'How much earlier?' Silas asks.

'An hour ago? No one's in here. It's not possible. I've checked.'

'Mind if we come in?' Silas asks.

'If that's what you need to do,' the technician says, stepping aside for them. Again, he keeps his eyes firmly fixed on the ground.

The first part of the brightly lit mortuary, the fridge room, is lined on one side with a bank of floor-to-ceiling refrigerator units. Beyond it is where the hospital post-mortems are carried out to establish medical causes of death. There are no coroner-requested autopsies done here, none of the investigations into violent or unexplained deaths that Silas has to attend, but the sight of the well-drained, stainless steel work surfaces, sinks and sluices has the same effect on Silas. This is a place of postscripts rather than prefaces, of deaths explained rather than prevented. For Silas, mortuaries will forever be synonymous with professional failure.

'Anyone else been down here this morning?' Silas asks.

'A consultant came in earlier,' the technician says, glancing at a worksheet on the counter.

Silas looks up. 'Who?'

The technician checks the worksheet. 'Dr Armitage – to sign off some cremation forms.'

Silas glances at Strover. 'And were you here with him?'

The technician stares at his feet. 'Most of the time. As I say, I went to fetch a tea at one point. I know Dr Armitage well. Everyone does.'

'But you went out when he was here?' Silas asks.

'I know him well,' the technician repeats. He turns away, visibly agitated now, eyes darting around the room but never at Silas. 'He's a highly respected consultant. Is there a problem with that? Me leaving him here on his own to fetch a brew?'

'You tell me.'

Silas walks over to the row of refrigerator doors, each one with a list from one to five on the outside with the names of the deceased kept within.

'What's the temperature in there?' Silas asks.

'These are positive temp cold chambers,' the technician says, his tone suggesting that Silas should know. 'Two to four degrees – cold enough to keep a body respectable for a few weeks.'

But not cold enough to kill someone. At that temperature, a person could hide in one for a while and survive.

'Are they full?' Silas asks. As part of the original contingency planning for Covid-19, the hospital had to build a temporary mortuary at the back of the staff car park for five hundred bodies.

'Twenty per cent,' the technician says. 'As I say, we're quiet at the moment.'

'Can we take a look?' Silas asks.

'Nobody could get themselves into one of these,' the technician says.

'To reassure ourselves?'

'And the bodies are all sealed up in heavy-duty bags,' the technician continues. 'You wouldn't be able to do one up from the inside either, if that's what you're thinking.'

'It should be easy to tell then, if someone's hiding,' Silas says, ignoring him. Everyone remembers the 'spy in the bag' case in Pimlico – except this man, it seems. The GCHQ employee, seconded to MI6, was found dead in a sealed-up North Face holdall. 'Can we start? Time isn't on our side.'

The technician turns to the first refrigerator. According to the list, there are two bodies inside. He's well built and swings open the heavy door with ease. There are three empty racks and bodies in black bags on the lowest two rungs. Silas glances at the technician, who finally looks up at him with tired, deep-set eyes. It's been a busy time for mortuaries.

'One arrived this morning, the other yesterday,' he says, turning away again. 'I put them in myself.'

'Check them,' Silas says. 'Please.'

Silas and Strover watch as he slides the body tray out by eighteen inches. He then unzips the top of the black bag to reveal a pallid complexion.

He reads out the man's name, his voice growing in confidence. 'Just as it says on the list.'

'Next,' Silas says, nodding at Strover. It was his dad who taught him to be thorough. Do something well or not at all.

They watch the technician repeat the process, sliding the body out with as much care and dignity as he can in the circumstances. Silas doesn't like asking him to do this, but he needs to find his man and there aren't many other places where he could be hiding.

After five minutes, they reach the final refrigerator unit. There is only one name on the list, in the lowest bay. The technician glances at Silas and swings open the heavy door. One body bag at the bottom of the fridge – and another on the level above. He closes the door a fraction, checking the list. Silas checks too. One name. Two bodies.

'That's not possible,' the technician says, blood draining from his already pale face. 'It can't be.'

Silas's mouth turns dry. Strover shifts on her feet, glancing at the uniforms behind her. They all look nervous.

'Bottom one first,' Silas says, watching the technician slide the tray out. He pauses for a second before unzipping the bag and then smiles in relief as he recognises the face. Swallowing hard, he slides the tray back in and puts his hands on the next one, checking with Silas, who gives the faintest nod.

Slowly, the tray slides out on its well-oiled wheels. Silas signals for the uniforms to be ready. Will the fake doctor make a run for it or accept that the game's up? Bracing themselves, they all watch in silence as the technician reaches over to the zip, fastened at the top. The technician's hand hovers above the bag. And then he pulls it open.

58

Bella

'Thanks, Gladys, I really appreciate it,' Bella says, heading west in her mum's car. Gladys had managed to stall the policeman long enough for Bella to drive away from the migrant centre.

'You should have seen his face when I dropped the pot,' Gladys says, laughing. 'The soup went all over them shiny black shoes he was wearing.'

Bella's grateful but she's in no mood to laugh. 'I've got to get off the phone,' she says, remembering Jim's words about not being tracked. 'Send me a text if Mum shows up.'

'Course, girl. I'm sure she will. But first you need to tell me wah gwaan with you and the feds.'

'It's nothing, really,' Bella says, thinking of Erin. 'I phoned the police with some information and now they want to speak to me. That's all.'

'So why don't you speak wi' them?'

Gladys is right. She should just go to the police. About Erin. Maybe her mum too if she doesn't show up.

'Long story – I've got to go. Thanks again, Gladys.'

Bella cuts Gladys off before she can ask any more awkward questions. Before the tears come. She doesn't trust herself to talk about Erin. It's as if a switch is flicked every time she thinks of her, diverting Bella into a parallel world. Her friend's death is just too extreme to be true. She tries her mum's mobile number again but it goes straight to voicemail. Same with the landline. It's so not like her. Unless she really is having an affair with Dr Haslam, in which case she might not be behaving rationally.

Bella turns on the radio, glancing in the rearview mirror. No one seems to have followed her out of London but she can't forget the man with jet-black hair who appeared in her street. The crop circle killings are still headline news, but there's no mention of Erin or the identity of any of the victims, which remains a mystery. Once again, there's an appeal for information – this time from amateur codebreakers. A group of academics is helping Wiltshire Police in their efforts to decipher the complex patterns. And a leading cereologist in California has claimed that the deaths are a warning from another civilisation about climate change. Whatever it takes.

She changes channels, thinking of the radio game. A road sign catches her eye. The next turning, in a mile, is to Salisbury... and to Porton Down. She remembers Jim's answerphone message from the pet shop at Porton. It's been haunting her ever since. Something about it wasn't quite right, the familiarity of the voice. Jim must be a good customer, but she seemed to know him better than that.

She glances at her watch. Jim had texted earlier, asking how she was doing, what time she would arrive in Swanage. She'd told him late afternoon. Another road sign to Porton Down. Taken on their own, the two words are innocent enough, but together they have an unsettling strangeness, forged over decades of dark rumours and conspiracy theories. She can't write a story about the place without a visit.

She pulls off the A303 and joins the A338 towards Salisbury. A quick detour, nothing more. She'll drive around the Porton Down site, get a feel for it. If the place really is involved in Erin's death, it's the least she can do for her. She'll also drop in on the pet shop, say that she's a friend of Jim's.

After driving through Cholderton and Allington, she turns off into the village of Porton. The pet shop is clearly signed but she carries on and is soon passing high-security wire fencing and the entrance to a long, tree-lined avenue. The Porton Down campus sign on the verge is familiar: the same as the one in the photo of Jim at his house. For a moment, she's tempted to drive down the approach road and ask at reception if she can talk to someone about the crop circles. They'd think she was mad.

She spins the car around and heads back to the pet shop, pulling up in the car park. It's better to start here, find her feet, reassure herself about the answerphone message. Two minutes later, she's walking around the aisles, looking at cute white rabbits, hamsters and big hairy spiders. She pauses in front of some guinea pigs before walking over to the reptile section to see if there are any lizards like Rocky.

'May I help you?' a voice says. She recognises it at once, from Jim's answerphone machine. *Hi, Jim, just checking to see if you're coming in to work today?*

'Having a look around, thanks,' Bella says, offering a

nervous smile of apology. The woman, in her forties, has kind eyes and is wearing a turquoise top.

'No worries, shout if I can be of any assistance.'

'Actually, there is something you might be able to help me with,' Bella says, as the woman starts to walk away.

'That's what we're here for,' she says, turning back to face Bella.

'I've got a friend who... comes here regularly, I think,' Bella says. 'He's got a chameleon called Rocky. And, well, I wanted to get him something for his birthday.'

'For Rocky or for your friend?'

Bella's sounding ridiculous, hasn't thought this one through.

'Both, really,' she splutters. 'You might know him, actually. I think he's a good customer. Jim. Jim Matthews.'

'Jim?' she says, her whole demeanour changing. 'Jim's not a customer. He works here. Should be on the tills now, in fact. Can't get hold of him. Don't know where he is, do you?'

59

Silas

Silas looks at the mortuary technician, his face drained of all colour, and steps forward to peer into the black body bag. The man staring back at him is not the fake doctor. He's not alive either.

'Recognise him?' Silas asks, but he already knows the answer.

It's a while before the technician replies. 'Dr Armitage,' he whispers, shaking his head in disbelief. 'I don't know how… I have no idea…' The technician bites his lip. 'He was such a good man.'

Silas whips out his phone and calls the estate manager, noticing a message from Mel to call her. Not now. Not when the body count's just risen to four and he's the SIO. Did the fake doctor lead them down here to buy himself time? After failing to be picked up at the front of the hospital, he entered the lower ground floor through the small staff entrance they came in by and walked straight out through the training academy, somehow avoiding the cameras. And what about earlier? He must have known that Dr Armitage was on his own in the mortuary. And then he killed him, put his body in the refrigerator and took Dr Armitage's security pass with him up to ICU.

'Did you see Dr Armitage leave here?' Silas asks the technician. The estate manager is not answering his phone.

'He'd gone by the time I came back from the canteen,' he says, looking down at the black body bag. 'At least, I thought he'd gone.'

The estate manager finally answers.

'It's DI Hart,' Silas says, glancing around the mortuary. 'You need to check the CCTV footage for the training academy, shortly after you saw the fake doctor coming into the lower ground floor. And search the footage for earlier too – for Dr Armitage.'

'We're still trying to locate him,' the estate manager says. 'His wife's really worried.'

Silas closes his eyes. Another person whose life is about to change for ever.

'We've just found his body in the mortuary refrigerator.'

Silas and Strover drive back to Gablecross police station in silence. Once again, a mortuary has left Silas feeling that he's come up short. The boss is not going to be happy.

'Any word from your boffins?' Silas asks, as he sits down in the Parade Room with Strover and glances at his phone. Three missed calls from Mel. It's her big day today. The fancy wedding. Lots of tuberoses. He'll ring her back later.

'Not yet,' Strover says. 'They're still convinced all three patterns loosely represent molecular structures – they're working through the Vigenère cipher codes.'

Silas sits back, contemplating the case, his options. The fake doctor has disappeared, along with the driver and passenger from the Range Rover. Forensics has yet to come up with anything from the vehicle. Samples have been sent for DNA analysis but there's a backlog, and Silas's repeated requests for fast-tracking have so far fallen on deaf ears.

All three people appear to be linked – the fake doctor signed the other two's hospital release forms – but Silas needs to establish a common motive. First the two men in the Range Rover visit Noah, trying to get him to reveal who had commissioned the crop circles. And then the fake doctor kills the third victim in hospital. If it was done to stop him talking, it would suggest that they thought he was the one who had commissioned Noah. What exactly did they fear would be revealed? And what is Jim Matthews's possible involvement in it all? According to Jim, the Range Rover tried to drive him off the road before visiting his house later the same day. And now he seems to have forced *them* to crash.

'Any more ANPR sightings of Jim's car?' Silas asks.

'Nothing,' Strover says.

Silas has taken his eye off Jim, the government scientist with an interest in mathematical crop circles. The two deaths at the hospital have become a priority, overshadowing all other lines of inquiry. But what if Silas is looking the wrong way and Jim holds the key? He should risk the boss's ire and contact Porton Down, find out what he can about Jim, report his conversation in the pub with Bella. And then there's Strover's hunch that Bella, who just happened to meet Jim in the pub, was the one who had called Crimestoppers with information about her friend Erin, the second victim.

'What do we actually know about Jim Matthews?' Silas asks.

They need to run the rule over him, if only to count him out. His gut feeling is that Jim's not a bad person. An oddball, yes, but not the sort to go around placing dead bodies in the centre of crop circles.

'Only what you found out in your interview,' Strover says. 'Have you rung MOD Police at Porton?'

'Not yet.'

Strover's silence says it all. She still thinks he should ring them. But she doesn't know their boss like he does.

'Want me to do some digging on the quiet?' Strover says, leaning towards her computer screen. 'See what I can turn up on him?'

Silas likes it when Strover gets her spade out. And he's learnt not to ask too many questions about her methods, how she uses Tor, the open source software, to trawl the depths of the Dark Web and contact her network of ethical hacker friends, the so-called white hats who perform SVAs – security vulnerability assessments – for companies.

'We just need to find out why the Range Rover was following him today,' Silas says. 'And why it might have tried to drive him off the road yesterday.'

'I could get Jim's tax records checked, see how much he earns,' she says. 'Maybe his bank account too. He might be in debt, susceptible to being blackmailed.'

It doesn't sound right, in keeping with the tidy, organised man he interviewed. 'Can we find out exactly what sort of work he's involved with at Porton Down?' he asks.

'Beyond what he lists on LinkedIn?' she says, looking at his entry on the networking site.

'I want to know more about his interest in BZ,' he says. 'What level of security clearance he has.'

'MOD Police could tell us,' Strover says, raising her eyebrows.

'And the boss will close us down like that' – Silas clicks his fingers – 'if he thinks we're making trouble, asking awkward questions. See what you can find out first. Maybe we haven't taken his meeting with the journalist in the pub seriously enough.'

Strover's phone starts to ring. Silas nods, indicating she should take it. And then his own phone starts to vibrate. It's Mel again. Maybe the tuberoses have wilted. He picks it up to answer and then puts it down again. Later. He'll call her later.

'Thanks,' Strover says, coming off her own call. 'Appreciate it.' She hangs up and turns to Silas. 'That was my friend in Hackney, a beat officer. He went to Bella's mum's place of work – a migrant centre in Homerton. Says they were definitely covering up something. As if they were protecting Bella from the police.'

'Maybe your hunch was a good one,' Silas says. He's still not convinced but he's learnt over the years that he's not always right. 'Do some digging on her too. Find out where she is now.'

Strover nods and then looks up at someone behind him. Silas turns to see one of the admin staff walking across the Parade Room. Everyone falls silent.

'Sir, I've got your wife on the line. Says it's urgent.'

60

Jim

Once he's sure that the Range Rover has gone, Jim packs up his laptop and slips out the back door of the old family home in Swanage. It's not safe to stay here, even if they didn't know for sure that he was hiding upstairs. They could return at any time. His plan is to head out of town on foot, take the South-West Coast Path up onto Ballard Down, where he used to walk when he was younger. It will give him space to order his thoughts, get some perspective on things. And it's safer than driving. If they followed him here, they will have clocked his number plate. At least no one's been to visit his dad. He rang the carer to check before he left the house.

He heads through the back streets of Swanage, up beyond the turning to the Grand Hotel, where his dad held his eightieth, and on past Ballard Down Stores, where Jim used to buy sweets as a young boy and cigarettes when he was older. He's about to take the battery out of his mobile when Bella calls. The news that it was her friend Erin in the second circle shocked him to the core, making it all so much more real. There must be a connection between Erin and Porton Down. Has to be.

'I was going to ring you,' Jim says, walking up between two houses where the town ends abruptly and the countryside begins. He's always loved this point on the route, the dramatic change in surroundings, like a border crossing. Its liminality. 'Are you OK? I'm so sorry about Erin.'

'Jim, I need to ask you something,' Bella says.

'Sure,' he replies, leaving behind the houses for open fields, scattered with sheep. 'Are you close?' It's strange talking to Bella as if they've known each other for years, but it also feels natural. And he can tell at once that she's not happy about something. Her voice is faltering, freighted with anxiety. It must have been horrific when she found out about Erin.

'Where exactly do you work?' she asks.

'Me? You know where I work,' Jim says, wrong-footed by the direct question. A couple has appeared on the steep footpath up ahead, coming off Ballard Down.

'I'm there now,' she says. 'In Porton, the village.'

'Why?' Jim asks. The MOD Police will challenge her if she's seen hanging around The Lab's perimeter fence. 'You need to be careful,' he continues. 'These people don't muck about, particularly with journos. And they followed me down to Swanage. If they think we're working together—'

'Jim,' she says, interrupting him. 'I'm not at the government site. I'm parked outside the pet shop. I've just been inside, talking to one of the staff.' She pauses. 'She said you work there – on the tills.'

Jim lets out a short laugh, thinking of the shop, how

much he enjoys its aisles, the range of exotic pets. 'Were you talking to Becky?'

'I don't know her bloody name, Jim, but you need to tell me what's going on.'

'Of course, I'm sorry,' Jim says, his voice quieter now. 'I should have explained.'

The path is steep and Jim has gained height quickly. He stops to look back on Swanage, spread out like a toy town beneath him. In the distance, a plume of smoke as the steam train carves its way through the valley towards Corfe. And then he hears its familiar whistle, carrying across the bay and down the years.

'Everyone who works at Porton Down must have a plausible cover story,' he says, turning to continue his ascent. 'In case someone approaches us, starts asking awkward questions about what we do. It's the first thing we're told when we get there. Get your cover story cleared with security. And it has to be something that's connected to our everyday lives, the closer to reality the better. My team leader's "legend", as it's called, is that he works as a barman at the nearby Porton Hotel. Which suits him down to the ground as he spends more time at the bar than he does in The Lab.'

'And yours is that you work at the pet shop in Porton?' Bella asks.

'I have been known to spend longer than my lunch break in there,' Jim says, his tone light and playful. He feels rejuvenated up here on the down, where kestrels hover on the sea winds. Invincible. 'Yemen chameleons are quite particular about their food. Nothing but gut-ready fresh locusts for Rocky! And I like the other animals they keep in the shop. It makes for a nice break from work. I find them very calming, they help me to switch off. Apart from the spiders. I'm terrified of spiders.'

It's a long time before Bella speaks. 'OK,' she says, hesitating. 'It's just that the woman said she was wondering where you were today,' she continues, 'why you weren't on the tills.'

'Well, now you know,' Jim says, stepping aside as a couple walk past him on the path.

'She also left a message on your answerphone,' she continues, 'when I was at the house, getting Rocky.'

Becky is nothing if not thorough – and unaware of the star role that she plays in Jim's cover story. She never will know. There's no need to tell her as his legend is sufficiently convincing, given how much time he spends at the shop in real life, how often he calls through with orders.

'And?' he asks. 'What did Becky want?'

'She was just asking where you were.'

It sounds as if Bella has lost momentum, abandoned her suspicions. She's not from a hostile foreign country, at least Jim hopes she's not, but it's gratifying that his cover story performed so well when challenged. Too well, perhaps. For a while, Bella genuinely seems to have doubted that he works at Porton Down.

'I'd ordered some locusts for Rocky and was meant to pop in and collect them today,' he says. 'Except that I was followed to work again by a Range Rover and came here, as you know.'

'OK,' she says tearfully. 'Should I still come? To Swanage?'

'Of course. But it's no longer safe in the town. We need to meet somewhere else, nearby.' He pauses, thinking through the best options. 'How about Studland Bay?'

It's in the direction he's heading anyway. Studland is an hour's brisk walk on the coast path that goes up and over the down rather than round past Old Harry Rocks, where he was thinking of going. As safe a place as any for them to meet. Busy, plenty of people. Jim and his dad used to walk there often. And the sooner he sees Bella the better.

'We could meet in one of the car parks,' Jim adds, worried by her silence. 'Maybe Knoll Beach.'

Once he's told her his story, shared his secrets about Porton Down, he will go to ground, take a ferry to the continent. He was intending to drive himself to Poole and ditch the car near the port, but she could drop him off after they've talked and she's seen the files.

'OK,' she says, her voice faint and weary.

'Call me when you're near,' Jim says, still troubled by her tone. 'You sure that's alright with you?'

'It's fine,' Bella says. 'Just that I haven't been to Studland Bay since my sister Helen left for Australia.' She pauses. 'It's time I went back.'

61

Silas

'How is he?' Silas asks, sitting down next to Mel.

'Not great,' Mel says. She's been crying and her eyes are bloodshot.

'Sorry I'm late,' Silas says, looking around at the bare white walls. They are in a consultation room at an acute mental health facility in Swindon. He puts a hand on Mel's thigh and she flinches, moves her leg away. It's taken him over an hour to get here since she rang the station. 'How was the wedding?'

'If you'd bothered to read your texts or answer your phone, you'd know it was a disaster,' she says. 'I had to leave early – to look after our son.'

'I'm sorry,' Silas repeats, full of remorse. Their son, Conor, is currently sedated in an adjacent ward, having had a psychotic episode in the middle of Swindon Old Town. They'd both thought he had turned his life around, after being diagnosed with schizophrenia two years ago, but it seems he's slipped.

'Apologies for keeping you,' a cheerful man in glasses

says, as he walks into the room and shakes their hands. Young enough to be fresh out of medical school, he introduces himself as Jonathan, a consultant psychiatrist, before running through Conor's current condition and medication with empathy and understanding.

'I dislike the umbrella term "schizophrenia",' Jonathan says, sitting back, hands clasped behind his head. His face is long, with a misshapen smile that suggests a state of permanent curiosity. 'It means "split mind" in Greek, which has led to all sorts of misconceptions and stigmas over the years – about the condition and its treatment. Your son – Conor – isn't suffering from a split or multiple personality. A shattered one, perhaps. And he may well have experienced psychosis, hearing voices – auditory hallucinations – as well as suffering from delusions. These are what, oddly, we call "positive" symptoms and typically they are controlled with drugs. More tricky to sort are the negative symptoms – the apathy, social withdrawal, poverty of speech, a general blunting of the emotions.'

'But he's been so much better recently, since coming off his medication,' Mel says, glancing at Silas for reassurance.

'Even got a girlfriend,' Silas adds. They both like Emma, how she's brought Conor out of his shell.

'Relapses are not uncommon after withdrawal,' Jonathan says, sitting forward to check his notes. 'Even if it's done slowly and carefully, as in Conor's case. It would be easy to put him straight back on the medication he was on, but there's a real opportunity here. I think it's important we consider carefully what medication we give him, and for how long.'

'What are our choices?' Silas asks. He likes this man, his

take on mental health. The faint whiff of theatricality that a lot of medics seem to have.

'Conor's choices,' Jonathan says, correcting him gently. 'One of the problems with long-term medication is that it can exacerbate certain symptoms, particularly the negative ones, as well as causing the brain to atrophy and shrink over time. And we can't dismiss what's known as tardive dyskinesia – uncontrollable, jerky movements of the face and body. Not such a risk with the newer meds, but we still see it. There's also the issue of weight gain to consider, a common side effect of many neuroleptics.'

'That's why we weaned him off his previous medication,' Mel says, turning to Silas, who nods. Conor had ballooned in size when he was being treated before. 'He's been getting into talking treatments, CBT and so on. We thought he was doing so well. Recovering. And now he's relapsed and we're back where he started.'

'It's OK,' Silas says, putting an arm around Mel, who has started to sob. This time she doesn't recoil from the physical contact.

'He *is* doing well,' Jonathan says. 'And a relapse following withdrawal of medication shouldn't be seen as a failure. He's on the long road to recovery. As I often tell my patients, it sometimes snows in May but summer always comes. In acute cases, medication is essential, of course, but my opinion is that Conor can make a full recovery without relying on them long term, once we've stabilised his condition.'

He glances at his notes, putting them down on the table as he sits back again. 'I'm a firm believer that schizophrenia, for want of a better word, is just one of many weird and wonderful ways of being human. And there's a responsibility on all of us to enable people like Conor to live as full and rich an existence as possible. I've certainly got no intention of wrapping him up in a chemical straitjacket for the rest of his life.'

Chemical straitjacket.

Time seems to stand still. Silas looks at Jonathan, studying his face in the silence as he plays back the words in his head.

'Sorry, what did you just say?' Silas asks.

He's conscious of Mel turning to look at him. Jonathan seems surprised and intrigued in equal measure, creasing his eyes as he scrutinises Silas.

'Just now,' Silas adds, one finger raised in the air as if he's trying to point to the exact moment in time.

'I think I said I've no intention of subjecting your son to a treatment that's akin to a medical straitjacket, as Thomas Szasz famously put it. Why?'

Silas doesn't know what to say or think.

'Szasz was a well-known Hungarian-born American psychiatrist,' Jonathan adds. 'Infamous for being anti-psychiatry when in fact he was simply opposed to coercive treatments.'

The consultant's words drift around Silas like wisps of smoke, twisting and turning in the claustrophobic room.

'You OK?' Mel asks, looking at him.

'I'm fine,' Silas says. He can't ring Strover now, not in the middle of a consultation about the welfare of his own bloody son. Not that Conor's condition will come as a surprise to the station. Conor was sectioned this afternoon by two of his colleagues in uniform after announcing that he was going to kill himself. Everyone will know by now.

'How else do people describe being on antipsychotics?' he asks.

Mel turns away, shaking her head in disbelief. The penny's dropped. She's realised that he's thinking about work rather than their son. He's there, at his wife's side in a time of crisis, but he's not there. A common refrain during the past year of counselling. And it's a valid complaint. His mind is already back in the Parade Room, making calls, pushing Strover, trying to progress the crop circle case and stop all the deaths.

'I'm not sure how else one would describe the appalling negative symptoms that can be induced by medication,' Jonathan says, still clearly puzzled by the detective. 'As I say, a general flattening of feelings and thoughts. Anhedonia, to give it its proper name. The inability to experience pleasure, or any emotional ups and downs for that matter.'

'Like a zombie, in effect?' Silas offers.

'Absolutely. In fact, I've heard that very word used by patients in the past.'

Silas nods, barely able to continue in the same vein. 'Almost as if they've had a lobotomy?'

Jonathan stares back at Silas, for the first time sensing another agenda. 'Yes,' he says. The penny's dropped with him as well. 'That too. Awful really, when you think about it.'

62

Bella

It's been three years since Bella was last at Studland Bay but it feels like yesterday. The place is crowded, even though it's late afternoon, but she manages to find a space in Knoll Beach car park. Why hasn't she been back here before now? Where has time gone? She thought she would be more nervous, frightened, but she feels calm as she turns off the engine.

She looks around the sandy car park, the bright orange pedalos lined up down by the shore, the sailing dinghies, the signpost saying it's 7,301 miles to Hawaii, 16 miles to the Isle of Wight. It feels like only yesterday that she was last here.

No sign of Jim yet. She'd called him fifteen minutes earlier, as she drove through Wareham. He said he'd be waiting for her here. The trip to the pet shop rattled her but Jim's explanation has made her feel better. She believes him, has no reason not to. He's clearly obsessed with Rocky, admits that he spends too long at the shop during his lunch breaks. What better cover story? The woman at the shop was certainly convincing. Did she think Bella was a foreign spy?

She opens the car door to cool down. Passing through Wareham had brought back happy memories of her dad. It was there that he'd boarded the steam train to Swanage with his two little girls, Bella and Helen, on one of their rare trips back to the UK from Mombasa. Her mum had driven over and picked them all up at the other end of the line in Swanage, meeting them with ice creams. A happy little family unit – until a year later, when he was shot in the head in Bakaara Market in Mogadishu. She wishes she knew where her mum was, why she'd left the house in such a hurry.

'Bella!' Jim calls out, jogging across the car park towards her.

She sits there, watching him approach with his big looping feet and childlike smile. For a second, she wonders if this is all a mistake, but then she smiles too. It's good to see him. Very good. Fate has thrown them together and she's not complaining. They're both loners, in their different ways. She gets out as he reaches her car and they hesitate awkwardly for a moment before hugging each other. Closing her eyes, she resists a sudden urge to cry. It feels nice to be held.

When she pulls back to look at him, maybe to kiss him too, she notices a nasty cut on his forehead. She'd forgotten about the night before, what she saw in his bedroom window, hoped she might have imagined it.

'What happened?' she asks, falling quiet, feeling a little dishonest.

'I knocked it, back at Dad's old house,' he says, trying to

make light of the injury. 'Forgot how low the ceilings were. Not the first time.'

She doesn't believe him, knows he's lying. Was he lying about the shop too? Jim raises a hand, feels the broken skin. 'I'm so sorry about your friend Erin.'

Bella casts her eyes downwards. Erin's death still hasn't sunk in, become a real event, something that she can look back on. Not yet. It seems to take so long for her to assimilate deaths into her life, to add them to her timeline. Many years, in her dad's case.

'A part of me still hopes it might not be her,' she says, shielding herself from the afternoon sun as he scans the car park, his eyes flitting from side to side. 'Is something wrong?' He's anxious again.

'I checked all the cars before you arrived,' he says. 'They don't seem to be here. Unless someone followed you?'

'Not that I know of,' she says. 'I kept looking, didn't see anyone. But the police turned up at my mum's work, where I was helping out. I left before they could ask me any questions.'

Jim shakes his head. 'The police in this country are unbelievable,' he says. 'I did warn you. Any word about her?'

'Nothing,' she says, tearful again. She so wants to trust this man who has stumbled awkwardly into her life.

They walk over to the ticket machine, where she pays for two hours. She has no idea if it will be enough, whether Jim's in the mood to share his story, help her to discover what might have happened to Erin. She senses they don't have long together.

'I wanted to talk to the police, ask them about Mum,'

she says, as they walk back to her car. 'But I couldn't trust them. Not after what you said. And that female detective who interviewed me last night?'

'What about her?' Jim asks conspiratorially.

'She rang back today and left a message. I'm scared, Jim. What these people did to Erin. You saying Porton Down's involved.'

'The police are up to their necks with MI5 in this,' Jim says, putting an arm around her. 'Trust me, you did the right thing by not talking to them. If they think I told you state secrets last night, they'll want to interview you as well as me. Bring us both in. You work for a national newspaper. You're a threat. Shall we walk down to the beach? We're safer in crowds.'

That's what her dad had thought too. Bakaara Market had been busier than usual that day, heaving with people buying everything from maize and fine gold to fake passports and petrol. But it didn't stop his killers. For so long, Bella refused to believe that he was part of the country's grim roll call of murdered journalists, but it was here on this beach that Helen had finally put her right, set the record straight about him.

63

Silas

'You've really got to make your mind up, Silas, think hard where your priorities lie in life,' Mel says, as they walk through the car park of the acute mental health facility in Swindon. 'Or this isn't going to work. You and me. Our marriage. Our family.'

'I'm sorry,' Silas says. 'I came as quickly as I could.'

'I'm not sure why you bothered, to be honest,' Mel says.

'Come on, that's harsh.'

'Is it? Is it really? When that nice consultant was going through the options – options that will decide the future quality of life for your own son – all you were thinking about was your bloody work.'

There's no point in arguing. She's right. The consultant had dropped a bombshell in the middle of his conversation with them, likening the effects of antipsychotics to a chemical straitjacket, and Silas had struggled to concentrate thereafter. The second victim, Erin, had been found tied up in one. Could the circumstances of all three crop circle victims relate to different medications?

He had tried to park the idea when they went to see their son afterwards. Conor was heavily sedated and didn't look well, but he'd managed to smile for them both, which made Mel cry again. It had brought a lump to Silas's throat too and both of them hugged him as he lay there between them in his hospital bed. Later, Emma joined them as well. It's her that gives Silas hope. She seems to understand their son, believes that he can live a meaningful life, one day perhaps without any meds.

'The case I'm involved in, it's taking a toll,' Silas says. 'I can't deny it. I was thinking about it when I fell asleep last night and it was the first thing that popped into my head this morning. It's eating me up. You've seen the papers, the news. It's caught the public's imagination. And they want results. Ward does too. So do I. Four people have died now and there could be more.'

'I know it's not easy,' Mel says, more conciliatory as she turns to open her car door. He'd driven over from the station in his own car and this is their only chance to talk.

'At least I left my phone in the glovebox,' he says, risking a smile.

'That's something, I suppose,' she says. 'Call me later.'

Silas watches as she accelerates away, lifting a hand in a half-hearted wave. He so wants to make it work with Mel, wants their son to get better. He also wants to crack the case – and he can't help feeling that the two are somehow related.

Five minutes later, he's chatting to Strover on the phone as he drives back to Gablecross. It feels strange not to be talking to her in person. She understood he had to leave the office for family reasons and she won't pry on his return.

They give each other privacy, their conversations never crossing the domestic threshold.

'Get on to your boffin friends and ask them to consider the possibility that these coded messages might have something to do with antipsychotics,' he says, driving through Swindon Old Town. 'And look up a famous psychiatrist called Thomas Szasz,' he adds, spelling the unusual surname for her. 'He once referred to these drugs as chemical straitjackets. See if he was talking about a particular one or just antipsychotics in general.'

'There are loads of them,' Strover says. He can tell she's already on the case, searching online as they speak. 'First generation, second generation.'

'Let me know what you find,' he says. 'Any news from your ring-round about missing frozen corpses?'

'Nothing,' Strover says.

It was always a long shot. There are hundreds of mortuaries in the UK registered by the Human Tissue Authority to carry out post-mortems. And then there are the hundreds of private mortuaries run by funeral directors. But mistakes do happen – the HTA recently recorded 250 serious incidents at hospital morgues over a three-year period, including wrong bodies being released to families and, in one case, two brains being mixed up.

'I've got to go,' he says, noticing an incoming call. 'Malcolm's on the other line.'

And Malcolm only ever calls when he has something important to say.

64

Bella

'It's just not like Mum,' Bella says, as they walk onto the sand and drop down to the water's edge. 'Leaving a note with no explanation.'

She looks around. Still no fear, no rising dread. She should have come back sooner. The beach is scattered with happy families on their summer holidays, sheltering behind bulging windbreaks, eating ice creams before they melt. Up by the dunes, a dreadlocked guy is flipping somersaults on a slack line, watched by a large crowd. And beyond him lie the beach huts. She'd forgotten about them. They were unsettling three years ago and they still send a shudder through Bella now. Something about the joyless, dark brown colours, the way the low-pitched roofs blend into the brooding pine trees behind them.

It was here where it happened. For a moment, Bella is eighteen again, reading Iain Banks on her towel in the sun, Helen lying next to her. This time, she does feel a twinge of something, a sudden twist in her stomach, like the snapping of a twig.

'You OK?' Jim asks, glancing at her.

She takes off her shoes and paddles in the shallow blue water.

'Maybe I should call the police,' she says. 'Report Mum missing.'

'Wait until I've gone if you do,' Jim says, taking his own shoes off to join her. His feet are pale and huge, like a pair of bloated fish. She holds out her hand and he takes it. 'They'll trace the call, work out where you are, who you are with.'

'Where are you going?' she asks, surprised by his sudden talk of leaving.

'Away from here, from them,' he says, looking out to sea. 'Maybe to France. I don't know yet.'

They keep on walking in the cool water, hand in hand, towards Middle Beach. It had all been going so well that day with Helen. She had made Bella laugh like a drain with her impersonation of two elderly male naturists – hands in the small of her back, pushing out her 'wrinkly tackle'. And then Helen had got mad with her about their dad and everything changed.

'Before you go, you need to tell me about Porton Down and the crop circles, how my friend ended up dead in one,' she says, keen to bring things back to the present.

'That's why I asked you down here,' he says quietly, slipping his hand out of hers.

It's the first hint of tension between them and she doesn't like it. She wants them to be friends, allies. Maybe something more.

'You know they found her body trussed up in a straitjacket?' she asks.

'I read that. I'm so sorry. Sometimes they use those for

the less willing "volunteers". I saw one with my own eyes – when I was at Harwell. If the crop patterns relate to different drugs that they test on humans, then Erin must have been a participant, willing or otherwise.'

Bella shakes her head. 'But she's never—'

'She might not have known, Bella.' His tone is urgent now, full of passion, as if he's possessed. 'Ronald Maddison died in agony as part of a research experiment in 1953. When they dripped a liquid onto his arm, he thought he was helping government research into the common cold – he had no idea it was sarin, a lethal nerve agent.'

Bella came across Maddison's case when she did some research at the pub last night. The only volunteer ever to have died at Porton Down. Apparently.

'It doesn't make his death any better,' she says, 'but at least Maddison knew he was being tested for something.'

'You know they'd offered him fifteen shillings to take part in the experiment,' Jim says. 'Money he was going to use to buy his girlfriend an engagement ring.'

They stop and look out to sea, watching as a pair of paddleboarders make their way down the coast towards Old Harry Rocks. On the horizon, dark clouds are rolling in from the Isle of Wight, bruised and threatening. She reaches out for his hand and holds it again.

'You see over there,' Jim says, pointing out to sea with his other hand. 'That's where a plane dispersed a large amount of toxic particles into the air, back in August 1959. Zinc cadmium sulphide. Part of a Cold War simulation to predict how an attack would play out if communist forces released a biological warfare agent off the south coast with a prevailing wind.'

'Are you serious?' Bella asks, looking back out to sea. It's hard to imagine on such a sunny day.

'They did loads of air and seaborne dispersion trials in the 1950s and early 60s,' Jim continues. 'Almost five thousand kilos of the stuff was released. The test in '59 seems to have gone wrong. For some reason, the Valetta aircraft dumped more than a hundred kilos in one go, ten miles south of Swanage. The sampling point at Dorchester recorded a whopping 4,300 particles. Places like East Lulworth were hit even harder as the cadmium cloud passed through on its way to Dorchester, but there's no record. Just a subsequent history of birth defects in the village. Cadmium gets into the lungs, kidneys and liver, you see. It's also carcinogenic.'

'That's awful,' Bella says, making a mental note to include the incident in her article.

'Par for the course for Porton. Are you certain Erin never volunteered for any drugs trials when she was at college?' Jim asks, glancing down the beach behind them again. His vigilance is making her nervous.

'Why would she do that?' she asks, looking behind them too.

'A lot of students do. I had a hard-up mate at Warwick who used to alternate between the sperm bank and phase one clinical trials – in between the odd lecture.'

'Not that I'm aware of,' Bella says, thinking back to the last time she saw Erin at college, how rough she had seemed.

'But you say she took a lot of recreational drugs,' Jim says.

Bella closes her eyes, pictures Erin stumbling into her college room, semi-conscious, mumbling incoherently.

'Sometimes I thought she had a death wish,' Bella says. 'She could be in a stupor for days. I didn't notice in my first year. We were all so out of it. But in recent months... It was like she wasn't there. Like she was a zombie.'

65

Silas

'Speak to me,' Silas says, taking the call from Malcolm as he pulls into the staff car park at Gablecross. Mel had every right to be angry with him at the hospital but he's pleased to be back on the case again.

'It appears your fake doctor injected his victim with a lethal overdose of anaesthetic,' Malcolm says, '500mg IV bolus of lidocaine, which led to hypotension, bradycardia, seizures and, sadly, cardiac arrest.'

'Sounds like he didn't stand a chance,' Silas says.

'The team fought hard to save him,' Malcolm says. 'But it turns out the lidocaine in his system, let alone the tetrodotoxin, was the least of his problems. His chest and abdomen were riddled with cancer – rapidly enlarging tumour masses on his lymph and spleen. I'd say he had a month to live at most. Maybe less.'

'Nothing to lose then,' Silas says.

'Exactly,' Malcolm says. 'Which has made me think more about who he might be.'

'Go on,' Silas says, turning up the volume on the call.

'Working on my theory that he's one of us,' Malcolm says, 'and your hunch that he was responsible for the condition, if not the death, of the other two victims, I've been looking at possible options.'

Silas nods, even though he's on the phone. He likes it when Malcolm gives him options. He'd like it more if he gave him simple answers, but Silas has learnt that's never going to happen. Malcolm's a forensic pathologist.

'He could be a mortuary technician,' Malcolm continues. 'But that doesn't narrow it down much. Or he could be a forensic pathologist. I know all the Home Office approved suspects – I used to arrange an annual dinner, in the days when we were valued for what we do – and he's not one of them. Or he might be a regular hospital pathologist. And this is where it gets interesting, given he carried out a lobotomy on the first victim and possibly infected himself with tetrodotoxin. Are you still with me?'

'I'm all ears,' Silas says, sitting up in his seat. He looks around the station car park, his car engine on to keep the air conditioning running.

'It's been nagging me ever since I saw the bruised eye sockets of the first victim,' Malcolm continues. 'A pathologist was struck off thirty years back for interacting unprofessionally with human tissue. I couldn't remember the details but I looked up the case and it seems that one of the charges against him, never proven, was that he'd performed a transorbital lobotomy on a deceased patient before carrying out a full autopsy. It was all a bit odd, given he subsequently removed the entire brain for examination. If a hospital colleague hadn't reported him, I'm not sure if it would have ever come to light, but it opened up a can of worms.'

'Where is this person now?' Silas asks.

'No idea. Struck off by the General Medical Council for a litany of other professional failings that were subsequently discovered. Most likely abroad, if he's still working as a pathologist. The Russians aren't fussy. Might be worth checking that key you found on him against all the morgues in Moscow.'

'Do you have a name?' Silas asks, wishing Malcolm wasn't so xenophobic. Silas once went to St Petersburg and found the Russian people to be welcoming and friendly. 'And does he look like our victim?' he adds.

'There's no photo, but he was called Steven Caldicott – that's Steven with a "v". And he used to work in the south. That's all I could find.'

'Is there any chance he could still be working in the UK?' Silas asks.

'Not officially, no.'

'Porton Down wouldn't employ him, for example, for some of their more dubious testing programmes?'

Malcolm lets out a long sigh. 'You really mustn't believe all you read about Porton Down, my friend.'

Silas doesn't, but it was worth a try. He looks around the car park. It's almost empty today. Everyone's on their summer holidays. He should be too, with Mel and Conor and Emma. Somewhere warm and sunny, with a medieval quarter that he can slip away to and explore when the beach gets boring.

'You didn't find anything else in his bloodstream, by any chance?' he asks, turning off the engine. He needs to get on, follow up on the forensics lead, brief Strover about Steven Caldicott.

'Like what?' Malcolm asks.

'Antipsychotic medication.'

'Not as far as I know. There were some unusual chemical traces in the first two victims – the toxicologist is still trying to establish what they are, but the fact that the bodies had been frozen for several months is making things more complicated.'

'Let me know if you find any meds,' Silas says.

'Can I ask why?'

'Just another hunch.'

66

Bella

'Can we sit down here?' Bella says, gesturing at a patch of sand up near the dunes on Middle Beach. 'I'm not feeling so good.'

'Sure,' Jim says. 'You look pale.'

'I'm fine,' Bella says, sitting down before her legs give way. It's all starting to come back to her, along with a rising nausea and dizziness. Where have these memories been hiding for the past three years? Why has she never confronted them until now, that day when Helen told her the truth about their dad?

'Here, have some water,' Jim says, passing her a bottle of Evian from his knapsack.

'Thanks,' she says and takes a sip. Jim squats on his haunches, unable to relax, eyes restless. 'Do you want to sit down?' she asks, patting the sand beside her. 'You're making me nervous.'

'I thought I saw someone, further down the beach,' he says, dropping onto the sand. He looks like a giant schoolboy, knees out, legs crossed. 'Sorry.'

'Was it someone?' she asks.

Jim checks the shoreline again. Bella follows his gaze. No one suspicious but the beach huts, dark and menacing, look as if they're watching them.

'Maybe, I'm not sure,' Jim says.

Bella's keen to talk to Jim about his story, what's on the USB stick, but she can't concentrate, not while she's on this beach. Maybe it was a mistake to meet here.

'Should we go back to the car, drive on to somewhere else?' Bella suggests.

'Here's as good as anywhere,' Jim says, smiling at her. 'You look better, more colour in your cheeks.'

She doesn't feel better. She feels worse. A lot worse.

'Can I talk to you about something?' she asks. 'Before you tell me your story.'

'Of course,' Jim says. He glances over at the car park. 'But we might not have long.'

'It's important,' she finds herself saying. Discovering what happened to Erin is important too, the reason she's here, but she realises there's something else she needs to deal with first, before she can move on.

'Tell me,' he says.

Bella leans back on her hands, scrunching the hot sand in her fingers so tightly that it hurts.

'We used to come here regularly,' she says. 'For family holidays.'

'Maybe we've met before?' Jim says cheerfully. 'I used to walk over this way when I was a boy.'

It's a strange thought, that they might have already encountered each other, given what's happened in the past twenty-four hours. 'On your own?' she asks.

'Just me. Sometimes with Dad. My mum died when I was very young.'

'I'm sorry,' Bella says, looking at him. From the moment they met in the pub, surprising things keep uniting them. Unspoken connections.

'The last time we were on this beach, I was lying over there,' Bella says, pointing at an area of sand below the dunes.

'All the family?'

'Me, Helen and Mum.' She pauses. 'I lost my dad when I was young.'

Jim nods. There's no need for him to say anything. The silence that falls softly between them is enough, full of mutual understanding.

'I now know that he died when I was eight,' Bella says. 'But I didn't believe it until I was eighteen. Don't tell me that's not weird.'

'Not weird at all,' Jim says, lying back in the sand on his elbows, more relaxed now. 'I thought my teddy could talk for worryingly longer than I should have done. What made you finally accept that he was dead?'

'Helen. My sister. For years she played along with it, my childish insistence that Dad had just gone away on a big story and would come back one day. It's what she'd told me, that first night after he'd died. When there was a power cut.'

She remembers it as if it were yesterday. In their shared bedroom at the top of the house in Mombasa, lit that night by candles. Bella was always spinning fantasies in her head, some based on the books she was reading, others pure make-believe. Her dad still being alive was just another story she told herself. He was lying low with the lions, moustache and beard scribbled on his face, investigating very bad people for the newspapers. Undercover for as long as it took to get the story.

'If you were only eight, I guess she was trying to be kind, help you sleep,' Jim says.

'Of course. Helen was the kindest sister. But it used to make her so vexed when we were older, that I was still waiting for him to come back from an assignment. The way I jumped up every time I heard the front door open, hoping he would scoop me up in his arms. One day' – she nods at the sand up by the dunes again – 'she lost it. Mum was swimming in the sea and I was asking Helen about Dad. Going on and on, desperate for snippets about him, anything to keep his memory alive. Helen was two years older than me and her recollections of him were so much clearer. I suppose it had become a family thing, that Bella the bookworm believed Dad would return one day, like the heroes in stories always do, but now I was eighteen and it wasn't cute any more. It was deeply disturbing. At least, that's what Helen thought.

'"Dad's dead, Bel, and he's never coming back," she said on the beach that day. "I can't believe that someone apparently so clever, always asking so many questions – so many annoying questions! – can be so fucking stupid. He's dead: D-E-A-D." I'd laughed, glanced out to sea where Mum was swimming and then back at Helen, who wasn't smiling. "You really don't get it, do you?" she said, shaking her head. My smile faded as her words echoed all around. *He's dead. He's dead. He's dead.* And then I...'

Jim looks up, encouraging her to continue. 'Then you what?'

Bella closes her eyes, tries to think back to what happened next, but all she can feel is the anger she had felt that day, building up inside her until it had risen to the surface like a panicked diver.

'What did you say to her?' Jim asks. 'To Helen?'

Bella's eyes spring open. She remembers. 'I didn't say anything.' She turns to look at Jim. 'I tried to kill her.'

67

Jim

Bella clasps her knees to her chest, tears streaming down her cheeks as she rocks backwards and forwards on the sand.

'Bella,' Jim says, leaning over to comfort her. 'It's OK.'

He wraps her up in his big arms, looking around him. A passing family stop and stare.

'It's fine,' he says to them over Bella's shoulder, wishing they'd mind their own business. He finds himself stroking Bella's hair, desperate to console her as she nestles under his chin. Wherever she came from, whoever she is, he hates to see her upset like this. They seem to have so much in common, their lives inextricably linked. Gradually, the sobbing subsides and they disentangle, the moment of intimacy over.

'I think we should walk back to the car,' he says, finding a hanky to wipe away her tears. The nosy family has moved on, but he's worried that she has drawn attention to them at a time when he was hoping to blend into the crowd.

'I'm so sorry,' Bella whispers, as they get up from the

sand. She's unsteady on her feet, like a newborn giraffe. He must watch *Peppa Pig* one day.

'It's OK,' Jim says, glancing down the beach. 'Do you feel able to walk?'

'I need to speak to Mum,' she says, pulling out her mobile phone. Jim watches as she turns it on and dials. It must have been devastating when the truth about her dad finally dawned.

'Still voicemail,' she says, looking up at Jim, as if he might be able to do something about it. He wishes he could. She's fighting to control her emotions, pressing her lips together as she tries to leave a coherent message. 'Mum, it's me. Can you call back? I really need to talk. Where are you?'

She hangs up and runs a hand through her hair, looking around her.

'It's OK,' Jim says, stepping forward to hug her again. They stand together for a few moments, her head resting on his shoulder. 'She'll call you soon,' he adds.

'I don't know what happened,' Bella says, as they set off slowly down the beach, his arm around her shoulders. He feels a gratifying arm slip around his waist too.

'Is this the first time you've been back here?' he asks.

She nods. 'I was so angry that day.'

'With Helen?'

'With her, with Mum, the world. It wasn't Helen's fault. I just didn't want to accept that Dad was never going to come back.'

Jim had felt real anger too when his dad was diagnosed with dementia, tried to deny it, despite the mounting evidence, the endless lists stuck to the inside of kitchen cupboards, the scribbled reminders on light-bulb boxes, the junk mail covered with Post-it notes. What had happened to Helen?

'When did your sister move to Australia?' he asks, deciding on an indirect approach.

'Soon afterwards,' Bella says, mopping her eyes with Jim's hanky.

'You'd been close up until then?' he asks, glancing at her. Bella's even more beautiful in profile, confident brow and strong, aquiline nose.

'Inseparable. We had our rows, of course, like any siblings, and she was always going on about me being the brainy one, which wasn't true, but there was never any serious beef between us.'

'But after the argument on the beach, she decided to—'

'Emigrate.' Bella lets out a short laugh. 'Pretty drastic, huh? We never talked before she went. I didn't get the chance to say sorry for what I'd done.'

'So was it serious?' Jim asks. 'What you did to her?'

Bella walks on in silence before she answers. 'Whenever I quizzed Mum about it later, she played it down, said it was a sibling squabble and would blow over. But it never has, so maybe it was more serious than I remember. Mum's still devastated.'

'There must have been something else going on in Helen's life,' Jim offers. 'To up sticks like that.'

'Maybe.'

They walk on past an extended family, spread across the beach like an army encampment, boundaries marked out with tartan windcheaters. Bella's eyes seem to linger on the checked pattern. Does it mean something particular to her?

There are coded messages everywhere, if only people know where to look.

'Me winning a place at Oxford didn't help,' Bella says, turning to Jim again, 'Helen had tried and failed. Several times.'

Jim had originally applied to Oxford to read maths – Dad thought he would have better job prospects – but after he was rejected, he went to Warwick and studied chemistry, his first love.

'Are you in touch with her now?' he asks.

'I send her letters, sometimes emails.' She pauses. 'And occasionally I call her, just to hear her voice, and leave long messages. She never picks up or replies.'

As an only child, Jim can't imagine what it's like to fall out with a sibling. He would love to have a brother, maybe a sister too.

They have come to a natural halt on the beach, standing in the late afternoon sunshine, summer holidays still playing out all around them. Someone is making sculptures, carving big circular patterns in the sand. Jim stares at them, tracing the swirling shapes. Is it another message? A warning? He looks up and down the beach before they set off again. He can't see anyone but something's not right. He can feel it in his bones. It's only when they're two hundred yards from Knoll Beach car park that he notices the Range Rover parked up in the shade of a tree.

68

Silas

'I want you to find out all you can about a pathologist called Steven Caldicott,' Silas says, sitting down next to Strover in the Parade Room.

He fills her in on everything that Malcolm told him in the car about the disgraced pathologist, and then turns to write 'Steven Caldicott' on a whiteboard on the wall behind. It's a new development, allowing boards on the walls in the Parade Room. Before today, they were only permitted in meeting rooms. These little things matter in office life. Like the day he put a bunch of Mel's leftover wedding flowers in a vase in the corner. Game-changing. Silas stands back, looking at the arrows linking Caldicott to the first two victims, wondering if he's the same person as 'zombie', written below. And then he wells up, suddenly thinking of Conor in the psychiatric ward.

'You OK, sir?' Strover asks.

'I'm fine,' Silas says, turning away, as if she might somehow be able to read his thoughts. He's never emotional in front of Strover. Never emotional in front of anyone.

Another thing that's come up in the counselling sessions with Mel.

'I've been talking to the boffins,' Strover says, thoughtfully changing the subject. 'About a possible link between the crop circle codes and antipsychotics. I rang them after you called.'

'And?' Silas says. Despite himself, he thinks again of Conor and his chest tightens. Mel's right. He should have paid more attention to the consultant, focused on his son rather than work.

'They think they're on to something,' Strover says.

'Go on,' he says, sitting down at his desk again. He wasn't lying when he told Mel the case was taking its toll. His body feels like it's been driven over by a tank.

'They've been running through the chemical formulas for every known antipsychotic,' Strover continues. 'First generation, dating from the 1950s, right up to the most recent, second generation atypical neuroleptics. Matched them against the crop circle's ASCII-generated letters and numbers and worked backwards, to see if they can break the code that way. No eureka moment, but one or two of the formulas for the most recent meds have got them interested.'

'What about chemical warfare agents?' Silas asks.

'That's the strange thing – they're getting similar results for BZ and VX. They're not ruling anything out at this stage. The molecular formulas could be medical or military.'

Silas turns to look out of the window. Should he drive back to see Conor? There's no point. He was so out of it, he hardly knew Silas and Mel were in the room with him.

'How about Thomas Szasz?' he asks. 'Discover anything interesting?'

'Quite a rebel,' she says. 'Believed that mental disorders aren't physical diseases – they can't be proved with biological evidence. He also had a thing against antipsychotics.'

'Chemical straitjackets.'

Strover nods. 'They dampen down hallucinations and delusions, but they…' She pauses, long enough for Silas to look up. 'They also extinguish someone's personality.'

This time it's Strover's turn to mask her emotions. She gets up from her desk and looks out of the window, her back to Silas. 'I had a friend who was on antipsychotics once,' Strover continues, her voice quieter now.

'I'm sorry,' Silas says. He's also surprised. It's not like Strover to reveal such personal information. Mel would be proud of them both. Does Strover know about Conor?

'Diagnosed with schizophrenia after one psychotic episode,' Strover continues. Silas can see her lean face reflected in the window, her tomboy hair and serious eyes. 'It wasn't a label she liked.'

'Schizophrenia?' Silas asks, remembering the psychiatrist's similar reservations.

Strover turns to face him, leaning back on the window ledge. She suddenly looks older than her twenty-eight years.

'People think you must be violent,' she says. 'And dangerous. And mad. People think you're nuts, basically. My friend wasn't any of those things. Just going through a rough patch. Things had built up, got out of hand. The drugs sorted her out – dealt with the immediate crisis, the hallucinations, delusions and paranoia.'

'The so-called positive symptoms,' Silas says. 'One of the great misnomers of our time.'

Strover nods. 'But she was kept on them for too long and the side effects were horrendous,' she continues. 'They seemed to exaggerate the negative symptoms. Emotional numbness, dry mouth, blurred vision, constipation, dizziness, weight gain. You name it. And don't even ask what it did to her sex life.'

'Is she still on them?' Silas asks, noticing that Strover's Bristol accent becomes more pronounced when she's emotional.

'Came off as soon as she realised what was happening to her, but that wasn't so easy either,' she says. 'She had a relapse but she's OK now. The sedation was the worst, the way it altered her personality. Annihilated it, more like. She couldn't help thinking afterwards that it was all about convenience – she was easier to manage in the community if she was a no one. If she didn't exist as a person.'

'If she was a zombie,' Silas says.

Strover looks up at him. Neither of them will forget last night, when they found the half-dead man on the hillside.

'I didn't make the connection,' Strover says. 'Until you called. Lobotomy, straitjacket, zombie.'

They both fall silent as she sits back down at her desk and Silas tries to concentrate on his emails. Strover eventually breaks the silence.

'I'm sorry about your son, sir,' she says, more formal now.

She knows. Of course she bloody does. It's why she told him all about her friend, how she recovered.

'Does everyone know?' he asks, his heart sinking as he glances around the Parade Room. He hates the thought of his family life being judged by competitive colleagues, who are only too eager to take advantage of another's weakness.

She nods. 'And everyone's hoping he'll make a full recovery.'

69

Bella

Jim puts his hand on Bella's to stop her walking any further along the crowded beach.

'What is it?' she asks.

He nods in the direction of the car park. They both watch as two men step out of a black Range Rover.

'Is it them?' Bella asks. They're taking an unnatural interest in her and Jim, pointing in their direction.

Jim nods, eyes locked on to the two men, who walk through from the car park and stop at the top of the beach, beside the water-sports hut. They are both wearing chinos and open-necked shirts, a gesture to the hot weather, but they still stand out in a sea of swimming costumes and bare flesh.

'You OK to run?' Jim says, looking around them.

'Which way?' she asks, her calf muscles tightening. His words have triggered a wave of adrenaline, washing away the strange feelings of earlier.

Jim looks behind them. 'Shit.'

Bella spins around to see another two men in the distance,

walking up from the far end of Middle Beach. One of them has jet-black hair and hunched shoulders – the man outside the newspaper office, her mum's house. And then Jim's tugging at her sleeve and nodding towards the dunes, where two more men have appeared out of the pine trees. They are moving less quickly, hanging back. One has his arm in a cotton sling, another appears to be wearing a neck brace.

'They must be the ones who followed me this morning,' Jim says, shaking his head. 'Had the argument with the tractor.'

'How about the sea?' Bella says. This is becoming all too real. Up until now, a part of her has been able to dismiss as fantasy the feeling that she's being followed. Even the policeman turning up at the migrant centre could have had an innocent explanation. Not any more.

'They wouldn't follow us, would they?' she adds. 'If we went into the water?' Already she can feel her body bracing itself for the cold of the sea. She hasn't swum since that day with Helen. They'd played Frisbee in the waves before the incident, like they always used to do, Helen flirting with some fit local boys from Poole.

'They can't touch us here,' Jim says. 'Not with this many people around.'

'So what do we do?' Bella asks. She's finding it hard to breathe.

'We keep walking and get into your car,' Jim says, setting off towards the car park. Bella hesitates for a second and then catches up with him.

'And if they stop us?' she asks.

'We say nothing,' he says. Jim's tone has changed, hardened. 'Unless they try to arrest us. In which case, we say a lot, as loud as we can. It's a crowded beach, people will be curious. And they won't like that.'

'Who are these men, Jim?' she asks.

'Does it matter?' he says, striding ahead. 'MI5, MOD Police? All I know is that they're here to stop me from telling the truth.'

Her too. They are a hundred yards from the car park now. The two men ahead have not moved from the wide open area in front of the café. They appear to be waiting for them. Bella glances behind, where the other two men are still following at a distance. Only the pair carrying injuries, up on the dunes, have kept their distance. Jim wasn't lying about the incident with the tractor.

'Jim, is this the right thing to be doing?' she asks, glancing at the sea again as they walk across the sand.

'We can't get away by swimming,' he says.

'Why not?'

'Because they've thought of that,' he says, nodding further down the bay, where a small speedboat is heading towards them.

70

Silas

Silas inhales on his cigarette and looks out over the fields around Gablecross. He shouldn't be smoking. Mel will kill him. But he needs a break after his chat with Strover. It was draining, all that emotional honesty. He feels rinsed out. Better for it, though. He can't deny it. Lighter.

'Sir, sorry to bother you.'

Silas looks up to see Strover in the doorway that leads out to the smoking shelter.

'Didn't know you smoked,' he says.

She humours him with a smirk. Strover's a runner, a fitness fanatic, wouldn't do anything to pollute her body.

'What is it?' he asks, stubbing out his cigarette. He doesn't like to smoke in public, not since he's given up.

'My mate in Hackney, he's been on the phone again,' Strover says. 'A woman who works at the local migrant centre in Homerton has been in touch with him. Not about Bella, but about her mum. My mate left his number with her when he visited the centre earlier. The mum's gone missing. Meant to be working at the centre today and failed to show up. Not like her at all, apparently. The woman had been around to her house and she wasn't there either.'

'And?' Silas says, still unconvinced. Strover continues to be fixated on Bella, convinced that she's a lead worth pursuing.

'I just think we need to talk to her,' she says, holding her ground. 'Ask if she rang Crimestoppers, that's all.'

'Because she mentioned she had a friend called Erin?' It's such a long shot, a small connection.

'There's something else too,' Strover says.

Silas looks up.

'Bella was arrested, three years ago.'

'Arrested?' Silas asks, thinking back to last night. Bella didn't strike him as the type to be in trouble with the law. 'You sure?'

'She was only eighteen,' she says. 'Family argument at Studland Bay. Breach of the peace. Just before she went up to Oxford.'

'Must have been some argument,' Silas says. Strover looks puzzled. 'I mean, more than about which flavour of ice cream to buy.'

'There weren't any more details,' Strover says, turning away. 'All charges dropped.'

Silas knows Studland well, took Mel there once for a weekend away. Spoilt her rotten at a local hotel called the Pig on the Beach. That was before Conor's problems, before their own.

'No mention of a friend called Erin?' Silas asks.

Strover shakes her head. 'I still think it might be significant.'

Despite his own reservations, Silas is pleased that Strover

has persisted with Bella as a line of inquiry. It's why they're a good team. Sometimes their hunches are different.

'Try talking to her again,' he says. 'And let me know if they find her mum.'

Silas expects Strover to leave him to his smoking, but she lingers in the doorway as he's about to light another cigarette.

'Something else?' he asks. 'Steven Caldicott?' He was expecting her to have come back by now with information on the struck-off pathologist.

'I'm working on him,' she says. 'It's Jim Matthews – I've drawn a complete blank, other than his time at Warwick.'

'That's not like you.'

He can't remember Strover ever coming up with nothing on someone.

'Porton Down's records aren't exactly the easiest to access,' she says. 'Military grade encryption, as you'd expect with a secret government facility.'

Silas knows what she wants him to do and lets out a sigh. 'I need to talk to the boss, don't I?'

She nods. 'It'll save us a lot of time.'

Strover's too polite to say how much time might have been wasted already.

'OK,' he says. 'I'll talk to him, ask if we can piss off Porton Down. He won't like it, you know that. Put us both back on cuticle duties.'

Despite Silas and Strover's best efforts, Swindon's nail bars continue to be targeted by human traffickers.

'That's the other thing, sir,' Strover continues. 'The boss has just been over, while you were out here. Wanted an update on the main suspects.'

Ward should mind his own business, let Silas get on with the case. He's got no right marching into the Parade Room demanding an update.

'I had to give him a quick briefing,' Strover says sheepishly.

'And mentioned Jim Matthews?'

'It was hard not to – his name was in big red letters on our shiny new whiteboard.'

With 'Porton Down' written in brackets below it. Silas can picture it now.

'He asked you to drop by,' Strover adds. 'Next time you're passing.'

Right now, in other words.

71

Bella

'How do you know it's them?' Bella asks, watching the approaching speedboat kill its speed as it nears a group of swimmers. The afternoon sun is still bright, but the sky to the west is streaked with amber.

'Trust me,' Jim says.

Maybe it's a mistake, impulsive, but she does trust him. With her life, if it came to it. She glances around at the holidaymakers, relaxing in their own worlds, unaware of the drama playing out in hers. Are there others hidden in the crowd too, watching, waiting? This is what happens when a government scientist tries to tell his story to a national newspaper journalist. She catches the eye of a middle-aged man reading the *Daily Mail*. Is he one of them? He folds the top of his paper down as Jim strides past. Bella offers a nervous smile. It's not returned.

'They might know where Mum is,' Bella says, as they draw to within fifty yards of the men ahead.

Jim turns to face Bella, putting his hands on her shoulders. 'Look, I know this isn't easy.' His breathing is fast, his blue eyes electric. 'If they take us now, you might never see your mum again. These people are dangerous – they tried to drive me off the road, ransacked my house, knocked me out. We're going to get into your car and drive away from here, head somewhere quieter. And then I'll tell you the whole story. If we do get separated, the password for the USB is 15071950 – my mum's date of birth. It's all on there.'

'OK,' Bella says, looking into his eyes. '15th July 1950.'

She leans forward and kisses him on the lips. Whatever happens next, she wants him to know how she feels, that she likes his big feet and Palmer glasses and awkward smile and pianist's fingers, but he seems surprised. Baffled even. Has she misread him, taken a liberty? His eyes widen and then he leans in to kiss her back, cupping her face in his tender hands. She wants the moment to last for ever.

'You've got the USB?' he asks, as a young child rushes over to retrieve a beach ball.

'It's in the car,' she says, smiling at the child, who squints up at them, standing too close.

Bella put the USB in the glovebox before she left London. Her laptop's hidden in the boot.

Jim looks across at the two men again. 'They won't want to cause a scene in public,' he says. 'So that's exactly what we'll do. Make a big bloody nuisance of ourselves.'

'Rant and rave like a couple of mad people,' Bella says, squeezing his hand. They're in this together now.

'Exactly. And if they do get heavy, I'll try to hold them back long enough for you to run to your car and drive off.'

'And leave you behind?' Bella asks. She'd assumed they would try to stay together.

'It's our only chance.'

Bella casts her eyes down. It's a desperate plan and she feels out of her depth, but he's right. There's no turning back now. A sudden thrill of fear passes through her. This is what she's always wanted to do. Break a big story. For Erin. Her dad.

Jim's already set off, striding towards the two men, who have deliberately blocked the path through to the car park. Bella slips a hand into her pocket and curls her fingers around the car key as she catches up with him. They have almost reached the men, one of whom fixes her brazenly in the eye. Who are these people? Have they families to go back to? Do they debrief with their wives? Sleep easily at night?

'Can we have a moment?' the man says, stepping in front of Jim, who tries to push past. Bella tucks in behind him like a celebrity dodging the paparazzi, but the man grabs Jim by the arm. Bella feels a hand on her shoulder too. The touch is very real. Before she can protest, Jim turns on them.

'Take your hands off me,' Jim says, shaking the man off as if he's contagious. His voice is loud enough for people to look up. Shocked by Jim's outburst, the men step back, discreetly letting go of their prey.

'We just need a quiet word,' the taller man says. 'Ask you both a few questions in private.'

'You're in my way,' Jim barks again, nodding encouragingly at Bella. She looks at the men and thinks of Erin. These are the people responsible for what happened to her friend. Erin would expect her to take a stand, go toe to toe with them. *Take no shit from anyone.*

'You heard what he said,' she says, remembering Jim's instructions to make a scene. 'Get out of our fucking way.'

She can't quite bring herself to shout but the force of her voice surprises her. Erin would be proud, particularly of the swearing.

'Listen, we don't want any trouble,' the man says, looking around him, assessing the unfolding scene. More people are taking an interest, coming out of the café and National Trust shop behind them. 'We just need to go somewhere quiet and talk.'

'Like a police cell, you mean,' Jim says, moving to walk towards the car park. 'I've heard they're nice and quiet.' Again, his path is blocked. Jim's taller than both men but they exude a strong physical presence. It's going to be a struggle for him to stop them long enough for her to make a run for it.

'You're not going anywhere until you've answered a few questions,' the other man says. His voice is quieter, more menacing. Bella glances around. A large crowd of people has now gathered to watch the altercation, arms folded, waiting to see what will happen. Several people have pulled out phones to film. Perhaps she should just go with these men, ask about Erin. Her mum. Dr Haslam.

Don't be so craven. It's her dad now. He loved words like 'craven'. And 'susurrus'. They crop up in the features he wrote about life in Africa – *the susurrus of the dry savanna.* She's read and reread them all in the past few months. She glances across at her car behind the men. Twenty-five yards away at most. She wasn't good at sprinting at school. Her long legs would get in a tangle in all the excitement. It was the cross-country runs over Hackney Marshes that she liked. Striding out on her own.

'Until *I've* answered a few questions?' Jim says, in

mock disbelief. 'How about you telling everyone here on the beach exactly what's going on at Porton Down.' Jim's raised his voice again, addressing the crowd like a politician at an impromptu rally. 'These men, they work for your government,' he continues, stabbing his finger in the direction of the man who grabbed his arm. 'And they don't want the truth to come out about one of the most secretive military facilities in the UK.'

'The ongoing programme of human experiments,' Bella adds. They'd make a good double act.

Jim turns to Bella, a mix of pride and fear on his face. And then his eyes widen a fraction. Time to run. But as her legs tense to sprint for the car, she hears one of the men say something to the other.

'*Folie à deux*. No question.'

72

Silas

'I thought I asked you to let me know if Porton Down ever became a formal line of inquiry,' Ward says. Silas turns away. He hasn't even been invited to sit down in the boss's office – always a bad sign.

'It's just one of many possible theories at this stage,' Silas says. 'We have no evidence to suggest a link between Porton Down and the crop circle killings and we haven't been making any inquiries—'

'Come on, Silas, we've known each other long enough,' Ward says. 'I saw it on your fancy new whiteboard. Who's this Porton employee you're so interested in? Jim someone?'

The boss is asking questions he already knows the answers to. Silas does it himself when he's interviewing suspects.

'Jim Matthews,' Silas says. 'A chemical analyst.'

'And have you put in a formal request for information about him?' Ward asks, steepling his fingers, as if he's just asked Silas to muse on the meaning of life. Silas often wonders if Ward wishes he'd gone into academia rather than the police. The bookshelves of his new office behind him are lined with tomes on religion and philosophy, a legacy of his days at Oxford, where he read theology.

'Not yet,' Silas says. 'I was going to ask you first.'

'Of course you were,' Ward says, smirking. 'After your own private inquiries had proved fruitless. You need to watch DC Strover. I had a call this morning about unorthodox computer activity from a Gablecross IP address.'

Silas reverts to his poker face, another tactic he uses when he's interviewing suspects. He's not going to rise to Ward's bait. And he's bluffing about Strover. She'd never be careless enough to use a Gablecross IP address for her online research, or a Virtual Private Network, however anonymous it might be. She's way smarter than that.

'What's so interesting about Matthews anyway?' Ward asks, getting up from his desk to walk around his office like a priest in prayer. It's hard to believe that Ward had to hot-desk it in the Parade Room with the rest of CID before his recent promotion to Detective Chief Superintendent. He's so bloody haughty.

'We were following another lead, in a village outside Marlborough, when we heard an altercation in a nearby house,' Silas says. 'We found Matthews covered in blood, claiming he'd been attacked after disturbing someone who was looking for classified material they thought he'd brought back from his place of work – Porton Down.'

Ward looks up, taken by surprise. 'And you didn't see fit to call this in? Sounds like a major breach of national security to me.'

Silas shakes his head. At least Ward doesn't seem to know about Jim's meeting with Bella the journalist. 'I decided to keep it in-house – on the basis he might be more relevant to our own ongoing inquiries into the crop circle killings.'

'We have procedures, Silas, which you've spent your entire career ignoring,' Ward says, sitting down at his desk again.

'Matthews has an interest in complex mathematical crop circles,' Silas continues, choosing to ignore Ward's jibe. 'The exact sort, in fact, in which the bodies were found. He also claimed to have had an altercation with a Range Rover, similar to the one we were already investigating in connection with the case.'

Ward nods in what most people would take as encouragement, but Silas is not falling for it. His boss is holding something back, a killer punch.

'And if his involvement in the crop circle case can be proved, Porton Down will be dragged into another public scandal,' Ward says. 'At a time when I explicitly told you we are trying to build bridges with the place after the Salisbury attack.'

'As I say, I was planning to clear it with you first.'

Ward sits back. 'I've got some good news, Silas.'

Here we go. Ward can be so patronising at times.

'Let me rephrase that. Good news for Wiltshire Police.' He pauses. 'Jim Matthews doesn't work at Porton Down. Never has done.'

Silas can't disguise his shock. He was ready for something, but not this.

'I don't follow,' Silas says, but he's already trying, tracking back over the evidence. They've only ever had Matthews's word and the results of an online search that threw up a brief entry on LinkedIn. Too brief.

'I made a call after I saw his name on your new whiteboard,' Ward says. 'This morning – while you were out of the office.' Ward lets the words hang in the air. When he was seeing Conor at the hospital. Does Ward know about his son's latest episode too? To be fair, he's always been sympathetic about Conor, which somehow makes it much worse. 'I came back to check with Strover, just to be sure. Matthews was due to start at Porton four years ago. Apparently they were very excited at his arrival. A gifted student, one of the brightest they'd recruited to their graduate development programme in years. He'd already done two summer holidays as an intern at Porton while at Warwick, and a job was ready and waiting for him on graduation. But he never showed up.'

'Where did he go?' Silas asks, still trying to make sense of it all.

'Nobody really knows. They were most intrigued when I told them you'd been in contact with him. They'd love to have him back. As I say, a truly gifted scientist.'

73

Bella

Bella hesitates for a split second, wondering what the man means by '*folie à deux*', and then she runs for her life, slipping away from the men on the beach like a sand eel. She feels a hand on her arm, but she's too gangly for them to get a proper grip. Running towards the car, she unlocks it remotely with the fob in her hand.

'I said get your fucking hands off me,' she hears Jim shout behind her. Has he managed to engage both men in a struggle? She hopes that she'll be able to thank him one day, for staying behind, giving her the chance to escape. Maybe he's trying to get some of the bystanders to join in? She hears heavy breathing, more shouts, grunts, the sickening thud of skin impacting on skin, but she daren't look back. She opens the car door and slides behind the wheel.

Her whole body is shaking as she fumbles with the key, trying to find the ignition. She risks a glance at Jim, who has managed to pin one man down, but his glasses are off, lying on the sand. He looks blind, like a giant mole in daylight. The other man is on his feet, wrestling to shake free from Jim, whose arm is linked around the man's upper leg. Jim's doing well, buying her precious time, but he can't do anything about the other two men now running up from the sand. They were behind them on the beach and are heading towards the car park. Towards her.

'Come on, start,' she says, turning the key in the ignition. Her mum's hopeless about the car, doesn't believe in getting it serviced. Jim's still stubbornly holding on to both men but the other two are now off the beach and less than twenty yards from Bella. Has she locked the doors? She feels for the key fob and presses it, just as one of the men reaches the car and slaps his hand on the roof to get her attention – as if he needs to. The next moment, he's trying her door.

'Get out of the car, Bella,' the man says, his face pressed against the window, contorted like a hideous gargoyle. How does he know her name?

She tries the ignition again, keeping her eyes averted. This time the car starts.

'And you get out my bloody way,' she says to herself, reversing into the middle of the car park. The other man throws himself in front of her, smacking the bonnet, but she slips the car into first gear, hits the horn and accelerates away. For a second, the man on the bonnet stares at her pleadingly, somehow hanging on, before he rolls off to one side.

Bella wants to be sick, but she knows she must keep going. Glancing in the rearview mirror, she sees the man get up from the ground and climb into a Range Rover that's pulled alongside him. She's got no chance against a car like that but she owes it to Jim to at least try to escape these people and tell his story. She reaches across and opens the glovebox, checking that the USB is still in there. 15th July 1950. What will happen if they catch up with her before she gets the chance to read it? Will they try to force her off the road, like they did with Jim?

She joins the main road and races away towards Wareham. A private ambulance passes by in the opposite direction, blue lights flashing. How badly have they hurt Jim? Checking her rearview mirror again, she watches as the ambulance turns down the road she's driven up and stops when it meets the Range Rover. The drivers start to chat. Bella slows up, still watching until they have disappeared around the bend in the road. Now's her chance.

She puts her foot down and drives on, not sure where she's going. Where can she go? It didn't feel safe in London. Back to Jim's house in the village? They will check there. She wipes away a tear and dials her mum's number on the car audio system. Voicemail.

'Mum, I'm in trouble here,' she says. 'I really need to speak to you. Where are you? What's going on?'

She hangs up and drives on towards Wareham, checking constantly in the mirror. What would her dad do? Change cars? Switch to another form of transport? She's got a bit of money in her bank account but not enough to stay on the run for long. She needs to open the USB, read its contents, find out what happened to Erin.

Jim did well to hold off the two men for as long as he did. She prays he's not badly hurt. He was kind on the beach, didn't judge her, listened about Helen. Gentle when he kissed her. They will take him away and charge him under the Official Secrets Act. At least, that's what Jim said would happen. Will she be charged too? She should ring her editor, tell him what's happened, ask him to clear the front page, expect a big story.

The Range Rover has still not shown up by the time she approaches Wareham. She drives over a bridge and up into the town, wondering why she's managed to escape so easily. It doesn't feel right. The Range Rover should have caught up with her by now. And then her phone rings. She'd accidentally left the battery in. It's an anonymous caller but Bella recognises her mum's voice at once.

'Bel, sweetie, stop the car and pull over,' she says, her voice barely a whisper. 'It's no good trying to escape the people who are following you.'

She sounds in pain, not herself. 'Mum, where are you?' Bella asks, glancing in the mirror. 'What's going on?'

'I can't talk now,' she says, her voice breaking with emotion. She's clearly speaking under duress, each word a struggle to get out.

'Are you OK?' Bella asks. 'What's going on? How do you know I'm being followed?'

Still no sign of the Range Rover behind her.

'I've got to go, Bel, my beautiful flower,' she says. Is she drunk? She hardly ever drinks. 'I'm so sorry I couldn't tell you.'

'Wait, Mum, what do you mean you couldn't tell me?' Is she with Dr Haslam? Is that what she means?

'Please Bel, just stop the car and wait for them to arrive. Do exactly as as they say and this will all soon be over.'

'What will be?' Bella's angry now. 'Christ, just tell me what will be over… Mum…? Please?'

The line has dropped.

74

Silas

'He might be covering up for Jim,' Strover says, as Silas spins restlessly in his chair in the Parade Room. He's already relayed the basic gist of what Ward said to her: that Jim Matthews apparently didn't turn up on his first day at Porton Down four years ago and hasn't been seen or heard of since.

'It's possible,' Silas replies. The same thought had crossed his mind. If Ward's trying to ingratiate himself with Porton Down, he might be doing them a favour by getting Swindon CID to drop an ongoing investigation into one of its employees.

'He definitely went to Warwick,' Strover says, searching for him again on her screen. 'And I know he's got no criminal record or major debts.' She shakes her head. 'The only direct refs to Porton Down are on his social media, including the entry on LinkedIn. If he's not currently an employee, he's making it all up.'

'Lying,' Silas says. 'Why would he do that? Pretend he's working at a place like Porton Down?'

'I'm sorry, I should have double-checked,' Strover says, blowing out her cheeks. 'Never trust social media.'

'And I should have contacted Porton Down,' Silas says. It's not Strover's fault – she'd asked him repeatedly to put in a call.

'I could request to check his PAYE records with HMRC,' she says. 'They won't specify Porton Down, but it will tell us if he's a government employee.'

'It's worth a try,' Silas says. Data protection has made it more problematic for HMRC to share personal tax information with other law enforcement agencies, but there are ways to ask – including ones that won't alert Ward.

'This email might change things,' Strover says, staring at her screen.

'Tell me,' Silas says, leaning over to take a look.

'Steven Caldicott.' She pauses. 'I did some digging. Seems like he moved to America after he was struck off by the General Medical Council. Started working for a small pharmaceutical company in Florida.' She looks up at Silas. 'Which specialises in antipsychotic meds.'

Silas leans in closer to Strover's laptop. 'We need a picture of him,' he says.

He watches as Strover's fingers move dexterously around the keyboard, typing so much faster than his own two-digit prodding. If she can find a photo of Steven Caldicott, they can match it against the face of the zombie victim. Silas is convinced that he placed the already dead bodies of the man and the woman in the first two crop circles, before taking tetrodotoxin and almost dying himself in the third circle. What he still doesn't know is who killed the first two victims. And when or why.

It takes five minutes of searching before they are looking at a thumbnail photograph of Steven Caldicott. It's an old image, taken while he was still a practising hospital pathologist. His GMC hearing was in the early 1990s – pre-internet – and there appears to be no reference to it, even on archived newspaper sites.

'I wouldn't exactly bet my mother's life on it being him,' Silas says, unable to disguise his disappointment. Maybe Malcolm's remembered wrongly. It's hard to compare the grainy image with the half-dead man Silas saw on the hillside and in the hospital. The pathologist in the photo is young and carefree, his smile a world away from the sallow lips of the zombie victim.

'Is there nothing more recent?' Silas asks.

'It's weird,' Strover says, still typing. 'Looks like he might have erased his online footprint, delisted himself from Google at some point.'

He and Mel had once tried to do something similar for Conor, after a string of teenage misdemeanours had affected his job prospects. It's not so easy.

'How did you find out about him moving to America, then?' Silas asks.

'A combination of Searx, Candle and DuckDuckGo,' she says, as if Silas should know what she's talking about. 'Dark Web search engines – not that it's really possible to trawl the Dark Web in the way that Google searches the regular internet. That's the point. I got my friend to do a trawl. Not really advisable for me to use a VPN and onion router in the office. She sent me the email. Said there was more to follow.'

Strover hasn't talked about her friend, the computer

expert, for a long while. Best not to ask too much – he'll just enjoy the fruits of her forbidden labour.

'There's more,' Strover says, switching windows to her email account, where a new message has landed. She clicks on an attachment and a photo of an older man stares back at them.

'Christ, that's him,' Silas says. A thrill of excitement runs through his ageing bones. It's moments like this that keep him going as a detective. The half-dead man on the hillside is looking alive and well. 'Do we know – does your friend know – anything else about him?' he asks, still studying the photo.

Strover switches to a second attachment. It's a screenshot of a company website listing small biographies – no photos – of senior members of staff. The third one down – Jed Lando – has been circled in red.

'Who the hell's Jed Lando?' Silas asks.

'Seems like Steven Caldicott's changed his name,' Strover says, scrolling to the next attachment, where her friend has typed up some brief explanatory notes. She flicks back to the second attachment. 'And retrained as a psychologist. At least, he says he has. Quite a career pivot – pathology to psychology.'

'What's the company?' Silas asks.

'Baylor Bristol Ottoman,' Strover says, opening up a new window on Google. 'Based in Florida, it's part of AP Brigham Inc.'

'Any UK connection?'

'AP Brigham owns a number of big pharma companies – and has recently started to invest in the UK.'

'Pharmaceuticals?'

Strover nods. 'And some health service providers too, by the look of it. Privately run, NHS-funded. Low- and medium-secure mental health facilities. Had a few run-ins with the Care Quality Commission over bullying staff behaviour. And it seems our Jed Lando was in charge of UK acquisitions and operations.'

75

Bella

Bella pulls into the train station in Wareham and drives down to the far end of the narrow car park that runs alongside the railway line. She hopes Jim's OK. A steam train is at the platform, waiting to leave for Swanage, a line of passengers queuing to board. Can she remember when she came here with her dad and Helen? She closes her eyes, calming herself down as she tries to picture the scene. Her hand in his as Helen skipped ahead down the platform. And then the terror of the train as it belched out a huge cloud of steam that engulfed Helen. Bella burst into tears, thinking her sister had been taken away, and hugged her long and hard when she re-emerged.

She wishes her dad were still alive, able to answer the questions swirling around in her head like the steam that day. What was wrong with her mum when she called just now? What was it that she couldn't tell her? And how the hell did she know Bella was being followed? It's still troubling her that the Range Rover seems not to have given pursuit. Is it because they know where she is going? She doesn't even know herself. And then she sits up, remembering something else about her dad.

For six months before he was killed, when he was alternating between homes in Mombasa and Mogadishu, he was worried that he was being followed everywhere by those who took exception to his articles on Somalia's brave human rights campaigners and peace activists. Each morning, before breakfast, he'd check his car for listening bugs and tracking devices, underneath and inside. He never found one, but what if something was attached to Bella's car in Studland while she and Jim were walking on the beach? It would explain why they're not in any hurry to follow her. They already know she's sitting here in the car park at Wareham station.

She gets out of the car, checks around her and kneels down on the warm tarmac, searching underneath the chassis. Nothing. She does the same around the passenger side and can't see anything there either. Their interest is scary but it means they are taking her seriously. They must know that she's been given classified information. But it still doesn't explain why she saw the man with jet-black hair outside the newspaper office in London and outside her house later. Both times were *before* she drove down to Wiltshire to meet Jim. What if she's been party to a sting, played an unwitting role in Jim's entrapment? In which case, her job is done. Jim's been caught leaking information to a national newspaper journalist and they don't need to follow her in a hurry. They'll just want the USB returned.

She's about to get back in the car when she decides to take one final look, underneath the boot. It's the place where her dad always checked last, before he came back into his flat in Mogadishu, rubbing his hands free of dirt to sit down for a breakfast of *canjeero* pancake bread and coffee. On the rare occasions the family visited, she used to peer under the car with him but it scared her as a young girl to see the axle and crankshaft and other nether regions of the car, as if they were dirty, out of bounds. And her mum didn't like it, because she thought that if there was a bomb, they might inadvertently trigger it.

Bella squats down beside the boot and looks underneath, just as the train lets out an angry hiss of steam. The sound makes her jump but not before her eye is caught by something small and rectangular, near the rear left tyre. She leans under and pulls firmly to release whatever it is. A container, no bigger than a match box, with two magnetic discs on one side, like miniature hot rings. Standing up, she looks around, digging her fingernails into her palms. The search was worth it. It also means she's a target.

The guard is signalling for the train to leave. She opens up the box and shakes out a small device nestled inside foam padding. On one side, four small symbols in white next to a row of tiny display lights: Wi-Fi, phone, power and battery life. It's a tracking device – and the power and phone lights are on.

The guard shouts for people to stand clear of the platform as the train starts to wheeze and spit its way forward, stirring like some ancient behemoth. Bella puts the device back in the box, an idea forming. She runs down through the car park, past the flower displays of bald car tyres brimming with bedding plants and onto the platform. The train has almost left but its last carriages are still in the station.

'Stand away!' the guard shouts, but Bella ignores him and

sprints along the platform, clasping the magnetic box in her hand. And then, as the hissing engine pulls away, Bella is alongside. She smacks the box against the metallic panelling of the final carriage and watches as the train heads off in a plume of smoke.

76

Jim

Jim doesn't know where he is being taken to or why MI5 has chosen to pick him up from the beach in a private ambulance. The interior is modern with glistening chrome surfaces and the strong smell of antiseptic. No doubt one of the many cover vehicles used by the security services. All he knows is that he can't move his arms or legs and his head is throbbing. He's short of breath too, his limbs still shaking with adrenaline after the fight on the beach.

'Where are you taking me?' Jim asks, his voice laced with raw fear. He's strapped into a wheelchair at the far end of the ambulance, its wheels bolted to the floor. His back is to the driver, and he can't move any part of his body. His ankles are fastened to the wheelchair and his arms have been secured across his chest in a white canvas straitjacket – just like the one he saw being used at Harwell. Without his glasses it's difficult to see, but two men are sitting on fold-down seats either side of him. He peers towards one and then the other and repeats his question.

'I said, where are you taking me?'

Jim's voice has no power, no conviction. Both men stare impassively ahead, as if they're sitting in the fuselage of a plane, waiting to parachute. There is no sign of a hospital stretcher or trolley, just a steel filing cabinet to his right. A large syringe has been strapped to its surface with a strip of masking tape – they had threatened Jim with it earlier, when he'd refused to get into the ambulance and wear the straitjacket. They must know that injections terrify him. There were too many at Harwell.

A phone starts to ring, breaking the silence. The man on Jim's left – shaved head, goatee beard – takes the call. He says nothing but glances occasionally in Jim's direction. After hanging up, he unbuckles himself and walks over.

'You should have come quietly,' he says, peering at Jim as if he's checking to see if anyone's at home. The man pulls out a small pencil torch and shines it into his eyes. Jim can smell garlic on his breath and turns his face away but the man grabs Jim's chin, forcing him to look forward again. There's nothing Jim can do. And he knows he should comply. It will only make things worse for Bella. He prays that she managed to get away and file her story.

'We've been trying to ask you some questions in recent days,' the man says, moving back to his seat, one hand braced against the ambulance ceiling for balance. 'But you haven't made it easy. We still need you to provide us with answers.'

Do they know about the USB stick? They must suspect that he's given Bella something, breached the Official Secrets Act that everyone has to sign at The Lab.

'I need to see a lawyer,' Jim says. He faces certain

imprisonment for what he's done and needs legal representation. 'Where are you taking me? Thames House?'

He walked past MI5's London headquarters once, on a stroll down the Thames with his dad, who pointed it out to him. He remembers the bars on the windows, how secure they were, a bit like the building at Harwell where he worked.

'We really need to know how you met Bella,' the man says, swaying to one side in his seat as the ambulance takes a corner too fast.

'Does it really matter now?' Jim asks, trying to make light of the question, but it's still troubling him too. Why *did* Bella come to his pub? And sit at *his* table? 'She happened to come along at the right time,' he says.

The two men exchange glances. 'Do you think someone sent her? To meet you?'

Jim manages a smirk. 'Of course.' Another look flashes between the two men like an electrical charge. 'Someone who clearly shares my concerns about Porton Down.' He pauses, flexing his ankles against the leather straps. 'What are you going to do with me? Lock me up and throw away the key? It's too late, you know. The story's already out.'

He's bluffing now, putting on a brave face. Has Bella managed to access all she needs on the USB?

'What do you know about the crop circles?'

Still the same man asking the questions. The other watches Jim with undisguised disdain. He was the one that Jim had hurt most on the beach. If Jim squints, he can just make out a patch of satisfying bruising above one eye.

'The patterns were very interesting,' Jim says. 'It's a while since I've seen anything so complex. But I had nothing to do with them. Or the bodies they found.'

He doesn't want to be charged with murder as well as leaking national secrets. And it's no bad thing if the state realises that there's more than one person unhappy with what's going on at The Lab.

'Clearly I'm not the only employee who's trying to draw attention to the abuses at Porton Down and its affiliated sites,' Jim continues, trying desperately to maintain an air of bravura. 'It will all come out eventually, you know. Just as it did with Ronald Maddison. You can't cover these things up for ever.'

The ambulance bumps over a pothole, throwing the two men around in their seats. Jim lurches in his wheelchair too but is unable to steady himself. The man with the goatee leans over and raps his knuckles on the steel partition behind Jim, expressing his disapproval to the driver.

'What did you tell DI Hart when he interviewed you?' he continues.

Jim's confused. Don't the police and MI5 talk to each other?

'I explained about Bella, how she'd approached me in the pub,' Jim says. 'And how your colleagues nearly made me crash on my drive home. We also discussed the crop circles – and Euler's Identity. It's a beautiful mathematical—'

'Did you talk about your time at… Harwell?' the man asks, interrupting him.

Jim shakes his head. 'I just said I worked at Porton Down.'

'Have you discussed your time there with anyone?'

'Only Bella. She's got the full story, I'm afraid. I've written a diary, you see. It's all in there.' Jim starts to feel stronger at the thought of someone else reading his account of what went on at Harwell. 'About my work. The experiments I conducted – and the ones I took part in.'

The man nods, as if he already knows.

A wave of fear passes over Jim as the magnitude of what he's done suddenly hits him. 'I'm in trouble, aren't I?' he asks quietly.

'Not necessarily.' The man pauses. 'Did you enjoy your time at Harwell?'

'Enjoy?' Jim asks, taken aback by the question. 'I'm not sure that's the word I'd use. Things got better towards the end. Why?'

'Because you're going back there.'

77

Silas

'I want to know everything we can about this Jed Lando,' Silas says, addressing a small team of detectives in a meeting room on the first floor of Gablecross. 'Where he lived, if he had family, his social media profile, who he met regularly and – most importantly – where he travelled for work.'

Silas nods at Strover, standing next to him. He's asked her to co-host the briefing for the first time. It's a big step up but he's keen to bring her on in his depressingly male-dominated team of detectives. 'We know he worked for AP Brigham Inc, a large American firm that was starting to run NHS-funded mental health facilities, mainly in the north-east,' she says. 'There are some further south too. And there might be ones we don't yet know about. They've been acquiring a lot of UK care groups in recent months, ever since the government allowed US investment in our health service.'

A murmur of disapproval rolls around the room like a Mexican wave. Silas was shocked to discover that US companies already provide almost 15 per cent of all mental health care inpatient beds in the UK. Even more surprised when Strover told him that the figure's as high as 60 per cent in her home town of Bristol.

'How can we be sure that Lando is responsible for placing the first two bodies in the crop circles?' someone asks.

Strover turns to Silas. Not unreasonably, given it's his hunch about Lando.

'There's no real evidence at this stage,' he says. 'What we do know is that Lando was a pathologist in his previous life in the UK before he apparently retrained as a psychologist in the States. The first two victims were frozen and one of them had been given a transorbital lobotomy, which suggests a person with at least some experience of a mortuary. Or possibly access to one.'

Strover glances at Silas before taking up the presentation again. 'As you know, we've yet to identify the first and second crop circle victims,' she says, eyes moving confidently around her colleagues. 'It's possible that both of them might have passed through one of these mental health care facilities at some time in their life.'

'Has there been any progress with breaking the codes of the crop circles?' someone else asks.

Silas nods at Strover, encouraging her to answer. She's been the one liaising with the eggheads on Zoom. Swindon team captain.

'We've had an update this afternoon,' she says, gaining in confidence all the time. 'It looks like the binary/ASCII code for each circle represents the molecular formula for a different chemical compound. We're not sure what they are – there are no direct matches yet – but we're currently working on two theories: either they're antipsychotic meds or chemical warfare agents.'

'Not messages from little green men, then,' another detective, the station joker, says. A ripple of laughter. Despite the deaths of four people, the crop circle element of the case has led to inevitable wisecracks at Silas and Strover's expense. Silas waits to see how Strover will respond.

'No surprise you're asking about little green men,' she says, turning on the joker, 'given you're on another planet most of the time.' More laughter, this time at the joker's expense. The noise quickly dies down, though, as the door to the meeting room opens and everyone looks up.

'Ignore me,' Ward says, slipping into the back of the room. The boss leans against the wall, arms folded.

Ignore him? As if everyone will just pretend that the most important man in the building hasn't joined their meeting. It's not a good sign. Silas uses the interruption to glance at his phone. It's been buzzing with messages throughout the meeting. Mel's worried about a change in Conor's care team at the psychiatric unit.

Jonathan, our nice psychiatrist, has vanished. When I asked, they said the whole place is under new management... American.

78

Bella

Bella watches the steam train pull away from Wareham station and jogs back to the car, ignoring the guard, who is still muttering about the dangers of running alongside moving carriages. He has no idea that she was attaching a tracking device to the locomotive's old metal bodywork. And she has no idea what the men in the Range Rover will think when they see the car tracker making its way down the narrow-gauge railway line to Swanage. She hopes her dad's smiling somewhere.

She hasn't got long, whatever her pursuers conclude. Has Jim put all the information she needs on the USB, including the link between the crop circles and Porton Down? She hopes he's OK. All she needs now is somewhere to read the contents of the USB – password 15071950 – and write up her story as quickly as she can, before emailing it through to the newspaper. They don't use copy-takers any more.

Bella drives north, through pine trees and isolated Purbeck countryside, looking for somewhere suitable to pull over. A sandy track into the woods catches her eye and she turns down it, parking where she can't be seen from the road. For a moment, she just sits, trying to process all that's happened. Jim's been taken away by people who've been following him – it now seems certain they're MI5. The same people were also following her, before she went down to Wiltshire on her newspaper assignment. Have they been following her mum too? She made no sense on the phone, telling her not to run away from these people. Where is she? She didn't sound herself. Should Bella ring the police? Tell them about the phone call, how distressed her mum sounded? Again, she hears Jim's words of warning.

She fetches her laptop from the boot of the car, takes the USB out of the glovebox and slides it in. An icon called 'Modern Maddison' appears, as before. Prompted for a password, she enters Jim's mum's date of birth with shaky fingers. It's so sad that Jim never really knew her. She can't imagine growing up without her mum. The password doesn't work. She must have made a mistake. Resisting the urge to panic, she glances around the deserted woods and tries again. This time she's in.

She breathes a sigh of relief, checking the rearview mirror. The track is deserted, the silence of the woods broken only by the sound of a distant car driving down the main road. The USB is divided into lots of different files – 'LSD', 'VX', 'BZ', 'Novichok Poisonings', 'Large Area Coverage Trials' – but one in particular, 'Harwell Diary', catches her eye. She clicks on it, remembering that Jim mentioned his three-year secondment to a secret unit at Harwell Science and Innovation Campus. In a short intro, Jim explains that he wrote the diary recently, based on recollections of his time there.

Some days are just a blur – we all worked so damned hard there – but other days come back to me out of the blue. I keep waking up in the night with a vivid memory of a particular incident or experiment and I write it down immediately, in case it goes.

Bella glances around the woods again and starts to read.

Week one

I am no longer on the Porton Down site in Wiltshire, but the spirit of that famous place is alive and well at this affiliated facility at Harwell in Oxfordshire, where I've been seconded. It's a relatively small unit – no more than thirty of us – and we're all here to help each other with proof-of-concept research into developing anti-dotes and countermeasures to emerging chemical and biological threats.

In keeping with the traditions of Porton Down, the on-site scientists at Harwell test a lot of the substances on themselves first, at less than lethal doses, of course. It's an approach that we, as new recruits, are encouraged to adopt from the moment we arrive. In many respects, it makes perfect sense, given that we are the ones who know exactly what chemicals we are working with. But it's still a shock when, on our first day, we are asked to queue up and are handed small phials of VX, one of the most lethal synthetic substances known to man.

Later in the week, I am injected with a low dose of a rare Russian incapacitating agent, a chemical derivative of fentanyl. It knocks me out for the rest of the day. Not everyone is so willing to participate. One of my fellow new recruits has to be physically restrained for the injection and placed in a straitjacket. He's screaming and shouting and almost bites clean through a senior scientist's finger. We are all too out of it to care but that night I fall asleep with an unfathomable sense of fear.

79

Silas

Silas feels sorry for Strover, making her first presentation in front of the boss. This is exactly what he feared would happen if he went to Ward about Jim Matthews. But he can't concentrate, not after the text from Mel. Jonathan, the psychiatrist they've been seeing, had given them hope and now he's gone. Mel's on the case, doing all she can for Conor – and still complaining that Silas isn't doing enough.

'What about Porton Down?' one of the team of detectives asks. 'There's been a lot of media speculation about a possible link.'

Silas winces at the mention of Porton Down and glances over to Ward, whose expression remains implacable. Did he get the question planted? Silas wouldn't put it past him.

'At this stage, we're not ruling out anything,' Silas says, pausing. He refuses to kowtow to the boss and pretend Porton Down isn't on the table. 'There remains a strong possibility that the crop circle codes might be related to military nerve and incapacitating agents, but right now I want you all to focus on Jed Lando. You should be aware though that Jim Matthews worked for two summers as an intern at Porton Down while at Warwick University.'

Silas glances over at Ward, still standing at the back, inscrutable as ever, and points at Jim's name on the flow chart behind him.

'Matthews seems to be of considerable interest to the driver and passenger of the Range Rover, who were linked to the fake doctor, who in turn we believe killed Jed Lando, the third victim, at the Great Western Hospital.'

He illustrates the complex chain of connections on the whiteboard, moving from one name to the next. 'What we don't know is why they wanted Jed Lando dead. Or indeed what their interest is in Jim Matthews, but it appears to be significant.'

After the meeting has ended, Silas walks downstairs to the Parade Room with Strover. The boss had left without comment when the discussion about Porton Down was over.

'You did well,' Silas says, sitting down.

'You think so?' Strover asks. 'I'm not sure the boss was impressed.'

'He had no right turning up unannounced like that,' Silas says.

'Was he there to put us off pursuing a Porton Down connection?'

'Without question. I'll look into Jed Lando, liaise with the team, you have another go at Jim Matthews.'

Silas can't shake off the feeling that Jim's more involved than they originally thought. Which means Porton Down might be too, whatever the boss thinks.

Twenty minutes later, Strover comes up trumps.

'HMRC have excelled themselves and replied promptly for once,' she says, turning to face Silas.

'And?'

'The boss is right. The last time Jim Matthews was listed as a civil servant was four years ago, when he was due to join Porton Down as part of its graduate development programme but never showed up – and was never paid. There are, though, several payments relating to two previous summer internships.'

'Anything since then?' Silas asks.

'Nothing for three years,' she says, reading from her screen. 'But get this: for the last few months, he's been working at a pet shop.'

'A pet shop?' Silas repeats, spitting out the words as if he's just swallowed a wasp.

Strover angles her laptop so Silas can see. 'Porton Garden Aquatic and Pets. I've looked it up on Google Maps – it's just around the corner from Porton Down, in the village.'

Silas stares at the screen in disbelief. It doesn't make any sense. Unless it's a cover story of some sort. Animal experiments are still carried out at Porton Down's laboratories, something that's not denied but not exactly shouted about either. Maybe Jim works in the animal labs, and has been given an alibi in case he's targeted by animal rights activists? It sounds unlikely but Silas has learnt that anything's possible when you're dealing with Porton Down. 'What was he doing in the three years before then?' Silas asks.

'Anyone's guess. No employment record – not one that involved paying tax, anyway.' Strover returns to her laptop and types. 'There's something else.'

Silas is still trying to get his head around Jim working at a pet shop. 'I came across a chemistry department interview Jim gave at Warwick uni after he won a prize,' Strover continues. 'He mentioned his parents, both academics. Sadly, his mum died when he was very young. Took her own life – in a secure psychiatric unit.'

Silas nods but he's thinking of Conor and of Mel's last text. A moment later, he's looking up the unit where Conor's staying and staring at a photo of the building – on AP Brigham's corporate website. It's the American company's latest UK acquisition.

80

Jim

'Why are you taking me back to Harwell?' Jim asks, looking from one man to the other in the back of the ambulance. It doesn't make any sense.

'Your work isn't finished there,' the man with the goatee says.

Another bump in the road forces both men to brace themselves in their seats. Jim falls to one side. He's still in a straitjacket, strapped into a wheelchair.

'It was a three-year deployment,' Jim continues, trying to figure out why MI5 wants him returned to Harwell. The Lab deploys a lot of scientists into the field to provide advice and analysis, most notably in support of military operations in Afghanistan. 'I'm not sure anyone could be there for much longer, to be honest,' Jim adds. 'It was hard work. Intense.'

For the first couple of years, Jim had put on a lot of weight, comfort eating to deal with the stress of life in The Lab. There's still so much he can't remember about his time there. The work was endless, days dissolving into weeks and months. Writing the diary has helped, but it's patchy, a series of snapshots.

'It seems you have unfinished business,' the man says.

'What do you mean?' Jim says, increasingly worried. 'I don't understand.'

Why is he being driven back to one of the government's most sensitive scientific facilities, having just leaked national secrets to a journalist? He was expecting to be taken into custody, not returned to Harwell. He glances again at the syringe on top of the filing cabinet. For a moment, he's back in the high-containment facility where a lot of the tests were conducted, where the air was thick with screams. 'Hold him down,' someone shouted. The young volunteer writhed on the floor, trying to break free from his straitjacket, his head twisting from side to side as the needle plunged into his thigh.

'Jim?' the man with the goatee asks. 'You OK?'

Jim looks around the ambulance and then stares at him, taking in his other features for the first time, the flat nose that looks as if it's been broken; the tobacco-stained teeth. Has he seen him somewhere before?

'Why am I going back to Harwell?' Jim repeats, holding eye contact. He's been doing well at Porton Down since his return, been a diligent employee. Apart from leaking classified information, of course.

This time, the man doesn't answer. There's no need to. Deep down, Jim knows. He's not returning to Harwell as a scientist. He's going back as a volunteer, a human guinea pig. It's the only explanation, what they do with renegade employees. And if he dies, no one will care. No one will know. It took fifty years for the truth to come out about Ronald Maddison.

He's back at Harwell again, on a bench outside the high-containment facility. He'd stepped away for a break from the intensity and secrecy of the work. This particular lab was biosafety level four and security was intense. How was he to know that breaks outdoors weren't allowed? Two MOD Police officers approached him, and then another two from the opposite direction. The place was fanatical about security, paranoid about intruders.

'Get back inside,' one of them barked. The younger guards were the worst. They were excitable, new to the job, eager to impress their colleagues, show off their training.

'I'm having a break,' Jim said.

'He's having a break,' another replied, his voice heavy with disdain. 'Did you hear that? You'll have a broken neck if you don't get back inside.'

Jim refused to move. Relations between the academic staff at Harwell and the MOD Police had deteriorated after one of Jim's colleagues had two vertebrae broken by heavy-handed security guards. A decision was taken by The Lab's scientists to draw a red line, not stand for any more abuse, which was why Jim decided not to move. Big mistake.

It took all four guards to drag him off the bench, but once Jim was on the ground, he could do little to stop them piling on top of him, like a rugby maul, kneeling on each of his limbs. They used to do that at school, until he was big enough to look after himself. This time there were too many of them. His face was pressed down into the soggy ground, a hand pushing on the back of his head until he could barely breathe.

'My glasses,' he tried to shout, but it was too late. They

were already off and disappearing beneath the shoe of one of his assailants, crushed and trodden deep into the mud.

He tried to fight back but they were too strong. And as they restrained him, twisting his limbs as if they were wringing them dry, and ignored his muffled cries for help, all the guards could do was talk about the summer – and what they had planned for their holidays.

81

Silas

Silas pulls up in the car park of AP Brigham's UK head-quarters outside Reading and turns to Strover. They are early for their meeting with the CEO, an American who seems to work weekends and was happy to see them this evening when he rang him. Silas didn't say what the meeting was about – it's always best to catch someone cold, see how they react to tricky questions. He was also worried that he might mention Conor, reveal that he's on their case as a father as well as a detective.

'Could murder a coffee, sir,' Strover says, nodding at a Starbucks across the car park.

Five minutes later, they are in the corner of the coffee bar. Strover hasn't made any more discoveries about Jim Matthews's life, but Silas is convinced that the young government scientist remains crucial to their investigation into the crop circle murders. He has also asked Strover to bring along the key that was found with Jed Lando on the hillside.

'We're missing something obvious about Jim,' Silas says, sipping on his double espresso.

'If the boss is covering for Porton Down and Jim does still work there,' Strover says, glancing at her phone, 'then HMRC are lying about his PAYE contributions and so is the pet shop.'

It seems unlikely. Strover had rung Porton Garden Aquatic and Pets on the drive over to Reading and was told that Jim Matthews was not only an employee but also in the doghouse. He had failed to show up for work today.

'The missing three years are what interest me,' Silas says. Jim just seems to have disappeared off the face of the earth between his aborted career at Porton Down and his apparent arrival at the pet shop.

'You really think Jim carried on working at Porton Down?' Strover asks.

'We can't rule it out,' Silas says. 'You have to keep an open mind when you're dealing with the funny brigade.'

'And if he was, what's the connection with the crop circles?' she asks.

Silas thinks back to his odd interview with Jim at his house in the village. 'Jim was convinced that the first pattern represented BZ, which we know is kept in small quantities at Porton Down.'

A few grams as laboratory stock, stored in the vaults for research purposes. It's an unsettling thought. Silas glances around the half-empty coffee shop. Nobody's within earshot.

'Let's assume, just for a second, that Jim's right about BZ and that he has continued to work for Porton Down in some capacity,' he says. 'We also know that he's into crop circles – complex mathematical ones. What if we've got this wrong and he was more involved in their creation than we originally thought? Not personally responsible for murdering the victims but maybe he helped to place them in the circles? What would his motive be?'

Strover looks away, lost in thought.

'He'd be a whistle-blower,' she says, turning back to face him. 'If the victims are also connected to Porton Down. A very cryptic whistle-blower, but that might make sense for someone like him.'

'It doesn't explain the pet shop, though,' Silas says, glancing at his watch. It's time for their meeting but Strover's phone has started to ring. He waits for her to finish.

'That was my friend in Hackney again,' she says, as they leave the coffee shop. Silas wonders if he's more than a friend. He's certainly gone the distance for Strover.

'There's been an incident at Studland Bay involving a young man and woman,' Strover continues. 'The police are in attendance. Someone took the number plate of a car that drove off at speed. It belongs to Bella's mum – that's why my mate called me. He also said there was another vehicle involved.'

'Another Range Rover?' Silas says.

Strover nods. 'I'm getting the full report sent over from Dorset Police in Poole. A private ambulance was in attendance too – blue-lit but not NHS-registered.'

It's a growing problem, not least in Swindon. Private vehicles have to be approved by the ambulance service in order to display a blue flashing light.

'Isn't Studland Bay where Bella was once arrested?' Silas asks, thinking back to what Strover had found out earlier.

'This time it seems she left before the police arrived,' she says.

'Who was the ambulance for then?'

'Tall man, big feet, thick glasses. Sound familiar?'

'We need to talk to Jim again.'

First, though, they have an appointment with Jed Lando's boss. And as they walk back over to the glass and steel entrance of AP Brigham's HQ, a black Range Rover with tinted windows pulls off the road and slides into a space in the company car park.

Definitely not a florist.

82

Bella

Bella sits back in the driver's seat and puffs out her cheeks. Her mum's car is warm, even though the windows are open, and she steps out for some fresh evening air. She wasn't expecting to read Jim's personal account of his time working at Harwell, of being trussed up in a straitjacket. When did he write it? It feels wrong to be intruding on something so personal, but he gave her the USB stick, so he must have meant her to read it.

She glances around the dark, deserted woodland, tenses at the sound of another passing vehicle on the main road, and gets back into the car. Dusk always makes her nervous. From what she has read so far, it seems that all ongoing chemical warfare experiments – the main angle of the story she will file – take place at Harwell in Oxfordshire, far removed from the Porton Down site in Wiltshire.

It makes sense, allowing the authorities to legitimately deny that human guinea pigs are subjected to chemical experiments at Porton Down, where the testing is solely for safety equipment and military clothing. But what about

Erin? Was she an unwitting human guinea pig at Harwell? It might explain the straitjacket she was found in. Except that the only volunteers are apparently the scientists themselves, who conduct experiments on each other.

She scrolls down to another entry and sits up.

Week three

Everything changes in my third week at Harwell. It's inevitable, I suppose – we can no longer keep subjecting each other to tests. There has to come a point when we increase our sample size and expose a wider section of the population to non-lethal doses of the chemical weapons we are working on.

Since the late 1960s, the government has insisted that Porton Down's work is about defence rather than offence. And a key part of that strategy is to study the effects of chemical weapons on humans. In the past few years, there have been more chemical attacks than ever before – in Homs, Kuala Lumpur airport and in Salisbury. The need to steel Britain's defences against them has never been greater.

My own focus is still incapacitating agents – 'humane' weapons that disorientate rather than kill, psychochemicals like BZ and LSD. Other scientists are assigned to work on lethal nerve agents – VX, sarin and novichok – as well as deadly, naturally occurring pathogens such as the Ebola and Marburg viruses. We all work in the high-containment facility, which at times has the feel of a hospital. White, featureless corridors, a strong smell of cleaning fluids, steel doors.

Incapacitating agents were first explored in depth at Porton Down in the 1960s and I'm charged with revisiting the data. Monday morning and I am given a young volunteer to work with – it's very much sold to us as a collaboration rather than an experiment. I'm not sure if the volunteer is aware of what he has signed up for, but I doubt it's being subjected to incapacitating agents. He's fresh out of university with a first-class chemistry degree and keen to get a toehold in the world of research science. He's probably on an internship, but we are encouraged not to bond too much with our allocated subjects, ask too many personal questions, and I soon discover why.

One of the key characteristics of incapacitating agents such as BZ, which mimic psychosis, is that they induce acute brain syndrome, characterised by delirium and delusions that can last for up to three days. There is also a lot of vomiting, staggering and stumbling, not to mention restlessness, dizziness, blurred vision, high temperature and tachycardia. My man experienced all of the above on our first day, in between bizarre, inappropriate laughter, dribbling stupors and bouts of irrational fear.

On Tuesday morning, I watch him in the test chamber from a reinforced viewing window in the door. He's a man about my age and it's horrific to look at him suffering, particularly as all I can do is observe and take notes on a clipboard. As I watch and write, he rips off his clothes

– disrobing is a common side effect of these drugs. Once naked, he has a look of utter perplexity on his flushed face, as if he can't understand the world, let alone his place within it.

At one point, when he is falling around the padded room, he stops to look over at me. We hold each other's gaze for twenty, maybe thirty seconds, every sinew in his body pleading with me to let him out. And then he runs headlong at the wall and knocks himself out. I put my hand to my mouth in shock and look around the corridor to see if anyone else has seen what's just happened. He's managed to find the one place where the wall padding is worn and smashed his forehead against it. A part of me can't help but admire him – he's found an escape from his torment.

83

Silas

Silas has learnt not to let personal prejudices interfere with professional judgements, but he's taken an instant dislike to the young American CEO of AP Brigham Inc. And he and Strover have only been talking to him for a couple of minutes in his shiny new office. It's not the thought that this man is ultimately responsible for Conor's welfare. Nor is it the pornographic posters of glossy pills on the walls, the electric guitar leaning against an amp in the corner, or the framed photo of him in what appears to be a rap video, dancing with an inflatable syringe. Nor the man's habit of smiling after each sentence, a grin so pronounced he's given himself dimples in his schoolboy cheeks. Smiling is a good thing – Mel's always telling him to try it more.

'So, guys,' – that's what it is: Strover and him being called guys – 'much as I love talking about the British weather, you need to tell me why I have the pleasure of your company this evening.'

Earlier, Silas had invited Strover to take the lead – he doesn't trust himself, now that there's a personal element to their investigation – but the look she gives him suggests she'd rather not. He doesn't blame her. It's the American's hoodie too. And the Adidas trainers. This man's CEO of a multinational company, not a bloody rock star.

'We're here about Jed Lando,' Silas says, watching for a twitch of the mouth, a touch of an earlobe – any telltale sign of a liar.

'Jed?' he says. 'He's left the company.'

The CEO's good if he is lying. Just not quite good enough. His body is still – too still. 'When did he leave?' Silas asks.

'Couple of weeks ago.' He looks from Silas to Strover and back at Silas again. 'Is there a problem, officer?'

Silas has broken a lot of bad news to people over the years, but this feels different, as if it will come as no surprise and bring little sorrow.

'I'm afraid Jed Lando died earlier this week,' Silas says, trying to sound sympathetic. As he suspected, he needn't have bothered. The CEO seems more concerned with brushing a speck of dirt off his chinos than by the news of a former colleague's death.

'I'm so sorry to hear that,' he says, struggling to sound sincere. 'Do we know what happened?'

'Not yet,' Silas says. 'Why did he leave the company?'

The CEO gets up from his desk and walks over to the window. 'Professional differences,' he says, his back to them. 'I'm not sure I'm allowed to say much more – there was some ongoing legal action between him and the board over the terms of his departure.'

'Your website says he's still on the board,' Strover says.

'Tell me about it,' he says, turning around. 'I've lost count of the times I've asked IT for a site-wide refresh.' He slumps back down at his desk again and swivels from side to side in his chair.

Silas could do with a site-wide refresh himself.

'Jeez, I'm so sorry to hear about Jed,' the CEO continues. 'Sure, we had our differences, but he was one of the good guys.'

'What exactly does your company do in the UK?' Silas asks, unable to cope with much more insincerity.

'That's one thing we do get right on the website,' he says. 'We provide a network of secure mental health services, run by us, funded by your wonderful National Health Service. And no, it's not an American takeover. Our sole mission, in collaboration with the NHS, is to progress service users, many of whom may be exhibiting challenging or dangerous behaviours, through a full-care pathway that's both therapy and recovery focused.'

Jesus wept. Is that what Conor is? A service user on a bloody pathway?

'You've also invested in several UK pharmaceutical companies?' Strover asks.

'Correct,' he says, reluctant to elaborate.

'Who manufacture antipsychotics,' Strover adds.

He looks at Strover long and hard, and then his eyes flick to the left. 'Among many other products, yes.'

'Does your company have its own morgues in the UK?' Silas asks.

'Morgues?' he repeats, unable to hide his surprise at the question. 'Some of our facilities are small secure wards within bigger hospitals, which I assume have their own mortuary facilities. I believe all our stand-alone secure units do have the necessary means to deal with any service user who passes. It's rare.'

'But it does happen,' Silas says.

He had asked Strover to check. More than fifty people who are detained under the Mental Health Act die every year in the UK.

'Sure, sometimes it happens. And every death's a personal tragedy. May I ask why this might be relevant to Jed Lando's passing?'

'What exactly was Jed's role in the company?' Silas asks, ignoring the CEO's question.

'That bit of the website was also right – he oversaw our UK expansion, commissioning and operationally managing facilities at regional director level, before becoming our Chief Operating Officer.'

'So he would regularly visit your various secure facilities around the country?' Silas asks – including Conor's, but he manages to say nothing.

'Correct,' the CEO says, glancing at his watch as he moves to stand up. 'I have a meeting in five, guys. Unless there are any questions that can't be answered by our website—'

Guys. 'Jed Lando was murdered,' Silas interrupts. The man hovers and sits down again.

'Murdered?'

Another poor attempt at surprise. Silas hopes Strover's getting this, filing it for future use. It's textbook stuff, top-drawer dissembling.

'Did he work from here?' Silas asks, looking around the office.

'He was based at another site,' the CEO says.

Silas tilts his head, encouraging the man to continue, but he says nothing.

'It wasn't on the website, that's all,' Silas prompts.

'He worked out of our medium-secure facility in Newcastle,' the CEO replies. His eyes flit to the left again.

'Where you have a number of facilities,' Silas says. The CEO nods.

'We might head up that way tonight, pay a visit,' Silas adds, glancing at Strover. AP Brigham began its UK investment in the north-east but Silas has no intention of driving there. Something's niggling in the back of his mind. He went through AP Brigham's website with Strover before they came here and he's sure the office contact number for Lando didn't begin with 0191 – Newcastle's dialling code.

'Now, I really must be going,' the CEO says, handing them both his business card as he gets up to leave. 'Call me if you have any other questions and I'll see what I can do. Except for the curveballs – maybe you could email me those first.'

'One last question,' Silas says, refusing to return the man's ridiculous smile. 'Whose is the Range Rover outside?'

The CEO doesn't miss a beat. 'That must be one of our courtesy cars – we use them to pick clients up from Heathrow. Like the one I'm late for a meeting with now,' he adds, glancing again at his watch.

Silas turns to Strover, who pulls out a photo from a wallet file on her lap.

'Is that one of your "courtesy cars"?' he asks, as the CEO takes the photo. It shows the Range Rover that was allegedly following Jim Matthews – after it had been involved in the collision with the tractor. The front is badly damaged, but the number plate is still visible.

'Don't think so,' he says, handing back the photo. 'All our cars have personalised plates – MFA1, MFA2 etc.'

'MFA?' Silas asks.

'Medication For All,' he says, smiling from ear to ear. 'It's our company motto.'

84

Bella

Bella can't read Jim's account of his secondment to Harwell quick enough. She knows she hasn't got much time left. The tracker will have arrived at Swanage station by now and her pursuers will have worked out what she did with it. They will be driving around the Purbeck countryside, scouring the lanes in search of her. Angry, humiliated by the trick she played on them. Her dad's trick.

She will write her story once she's finished reading through all the documents. It's the least she can do for Jim. The least she can do for Erin. Bella doesn't know how or when it happened – during uni holidays? – but her friend must have been a volunteer at Harwell, where scientists subjected her to experiments with incapacitating agents. According to Jim, the particular pattern of the crop circle she was found in depicted LSD.

She reads on.

Week seven

As part of our daily work, we are told to revisit the US military research that was conducted on BZ, as well as the UK's own experiments at Porton Down, which means sifting through the records from Edgewood Arsenal, a former secret facility on Chesapeake Bay run by the US Army Chemical Corps in the 1960s. Under the guidance of Colonel Jim S Ketchum, a psychiatrist, some of the thousands of volunteers who passed through Edgewood were subjected to psychochemical experiments involving LSD and an anticholinergic, 3-Quinuclidinyl benzilate, otherwise known as BZ.

My own human guinea pig, who has already been run pretty ragged after a month of trials, does not respond well when I increase the doses of BZ. Originally rejected as an ulcer therapy, it can induce psychosis, followed by amnesia, although I don't think either of us will ever forget the night that brought an end to my own particular programme of experiments.

My task is to recreate the Boomer, a 1960s CIA variant of BZ that can can trigger a sustained, two-week period of delirium and acute mental disorder. After checking the padding on the walls is intact – my volunteer has a habit of ripping it off and stuffing pieces into his mouth – I signal for one of the staff to release a cloud of the gas into the cell from the usual aerosol dispenser and we leave him alone.

He gives me that pleading look again and starts to fidget and agitate, picking at the bare mattress on the bed, slowly at first, as if he is weeding daisies from a lawn, and then more violently, clawing at the material. And all the time he is mumbling – I can hear some of it relayed on the loudspeaker in the corridor outside, but it's mostly gibberish.

At one point, he stands up on the bed, looks below him and starts to shout, jumping up and down. He seems to be hallucinating, talking of giant spiders crawling all over the floors and walls, hundreds of them. And then his head begins to twitch spastically, from one side to the other. His hands flap against his chest and face, as if he is on fire and trying to dampen the flames. I can't watch any longer and call for help.

By the time Dr Haslam arrives, the young man is lying motionless on the floor. He has a temperature of 103.6 and we all fear the worst. But Dr Haslam is reassuring, talks of the dangers of puncturing delusions. And sure enough, two days later, after being given an antidote, the man is as right as rain.

85

Silas

Silas and Strover walk across the car park towards the Range Rover with its MFA personalised number plates. Silas managed to bite his tongue when the CEO explained that the letters stood for Medication For All, didn't take issue with him about AP Brigham's prohibitive pricing strategies, the implication that everyone needs meds.

'Put money on him watching us from his office window,' Silas says. 'Don't turn around,' he adds, sensing that Strover is about to check. 'Just pass me the folder.'

Strover hands over the folder and Silas drops it at the back of the Range Rover, the photos spilling out on the tarmac.

'I'll get them,' he says, as she moves to pick up the photos. He leans down and gathers them up, but his only interest is the number plate, which is now at eye level. The fittings are new. He glances up at the CEO's office as he gets back to his feet. Sure enough, the man is watching them. Not in such a hurry for his meeting after all.

'The rear plate's been replaced recently,' Silas says, back

beside Strover as they walk over to their own car. 'Find out how many courtesy vehicles are registered to the company.'

Forensics is still trying to establish the legal owner of the impounded Range Rover that was involved in the incident with the tractor. As Silas suspected, its show number plates are illegal – they are registered to an unwitting Range Rover owner in Birmingham, who wasn't happy to discover that his car had been cloned. The chassis and Vehicle Identification Number have also been restamped. It's a ringer, in other words, making it impossible to trace the owner. A ghost car.

'I hope you were taking notes,' Silas says, as they drive away. 'How to spot a liar.'

'And there I was, thinking someone's innocent until proved guilty,' Strover says, smiling.

'Unless he's the boss of a dodgy big pharma company,' Silas says, resisting the temptation to say something about Conor. He's told Strover of the connection and she was as shocked as him. 'He was also right-handed and kept looking to his left,' Silas adds, recalling something his dad once told him. 'Right-handed liars usually look to the right when they're recalling facts, to the left when they're bullshitting.'

It's by no means a foolproof method but Silas is surprised how often it works.

'What if you're a leftie, like me?' Strover asks, glancing at him.

'You look to your right when you're lying. Our CEO wore a watch on his left wrist. More to the point, he also hiked the prices of three essential medicines in the US by a factor of fifty – yes, fifty – when he bought them last year,' he adds, still incensed by the encounter, the number plates. '"Medication For All", my arse. And they've had nothing to do with manufacturing Covid-19 vaccines or anything useful like that. I also never trust anyone over the age of forty who wears a hoodie to work.'

'He's only just turned forty,' Strover says.

Mel bought him a hoodie once, in an effort to refresh his middle-aged weekend wardrobe.

'And he didn't seem in the least bit surprised by Jed Lando's death,' Silas says. 'He was definitely trying to throw us off the scent by mentioning Newcastle.'

'Are we really driving there now?' Strover asks, glancing at her watch.

'Not if we can help it. Check Lando's contact details on their website for me, will you? I'm sure it wasn't a Newcastle number.'

Strover looks it up on her phone. '01865 – Oxford,' she says, scrolling through the website. 'Where they've got a medium-secure psychiatric hospital – Cranham Hall.'

'We need to get over there now, find out if Lando was based at the site, talk to some of his colleagues,' Silas says, flicking on the blue lights of their unmarked car. 'And see if the place has any mortuary facilities.'

86

Bella

Bella reads and rereads Jim's diary entry, trying to ignore a sickening feeling in her stomach. 'Dr Haslam' is an unusual name. Maybe it's a coincidence? But she knows that it's her tutor at college – and that Erin is somehow involved.

If her friend ended up dead because she was part of some hideous experiment at Porton Down, or its affiliated site at Harwell, then Dr Haslam is the missing link that connects them. Harwell isn't so far from Oxford – sixteen miles. But what the hell would her English tutor be doing at a government science park? Maybe he's on some sort of retainer, siphoning off sick students to become human guinea pigs, and came to visit. It doesn't add up.

She knows she should drive off, keep moving, but she can't resist one more look at the diary, to see if Dr Haslam features again. Searching for his name, she is taken to an entry near the end of the document.

Year three, week forty-seven

It's time for my final assessment with Dr Haslam. He's pleased with my three years at Harwell, and full of optimism for the future. My concluding experiment involved me being given an injection of a slow-release incapacitating agent from China. Only a few of us were chosen for the test and I'm glad I was one of them, despite having to endure another needle. For once, the side effects were minimal and we've all been scratching our heads, wondering how the Chinese expect to incapacitate the enemy with something so mild. It's a high note on which to finish.

'There comes a point in every person's studies when things fall into place and it's time to move on,' Dr Haslam begins. 'Time to go out into the world and put science into practice. In your case, you'll be transferred back to Porton, as you've requested, to work in the labs – a short rehab stint with the animals. I know you are something of an animal lover and are keen to ensure welfare standards are maintained over there. It might seem like a sideways move, but it's anything but. We are very conscious of the need to rehabilitate scientists after such an intense secondment here. You've also had a fantastic few months – you've lost weight, look in good shape, you're thinking clearly, working well with colleagues. It's what I like to call an "awakening".'

Bella stares at her laptop screen, open-mouthed. It's the same Dr Haslam, using the same language as he used when she left Oxford. She had her own 'awakening' too, an epiphany. The day when the fog finally lifted in one of their last one-to-one tutorials and she caught a glimpse of life beyond, a way forward. Her hands are trembling.

She looks around the woods, where the tall pine trees have started to spin, and opens the car door for some fresh air. Why's Dr Haslam overseeing Jim's work as a government scientist at a secret facility in Harwell? And did he play a role, however unwittingly, in Erin's death? She needs to speak with him, get him on record for her article.

Should she read more of what's on the USB or head straight for Oxford, confront Dr Haslam? He might still be in Italy but she feels powerless sitting in her mum's car. The Range Rover will be looking for her and she has a sudden urge to get as far away from here as she can.

87

Silas

Silas and Strover drive on towards Cranham Hall at speed and in silence. The journey should take less than forty minutes. And they might catch the staff unawares, if the CEO bought his story of driving up to Newcastle and hasn't warned them to expect a visit from the police.

'Did you see that photo of the CEO in a rap video?' Strover asks, scrolling down through YouTube on her phone.

'Unfortunately,' Silas says.

'It caused quite a stir in the US last year – they showed it at their national sales conference, hoping to persuade more GPs to prescribe their meds,' Strover says, still glued to her phone. 'Another pharma company did a rap video promoting their brand of fentanyl, a synthetic opioid. One of the execs was dressed as a giant spray dispenser. He's now been found guilty of racketeering and bribing doctors to prescribe their product.'

Strover starts to play the AP Brigham video on her phone. Silas has never been a big fan of American rap music – old-school hip-hop is more his thing – but there is something truly dire about the song now playing. And then it gets even worse: a character dressed as a syringe joins the rappers.

'What the hell?' Silas says, glancing at the dancing needle. It looks like a cross between Teletubbies and a drill video.

'It's Brigham's latest med,' Strover says, reading from her phone. 'An antipsychotic depot injection – slow release, slow acting. Perfect for people who forget to pop their pills. You only need to take it once a month. Once every six months, in this case – it's a game changer, apparently. And much more expensive than daily meds, of course. Six grand a year.'

Silas is about to ask her to tell him more – Conor was terrible at remembering to take his pills – when his phone rings. It's Malcolm.

'You still interested in Caldicott?' he asks, forgoing any formalities. Strover kills the video.

'More than ever,' Silas says. He'd rung Malcolm back after they'd identified Caldicott as the third victim, asked him if he could discover anything else about his past career as a pathologist.

'There's me having a go at you for being a conspiracy theorist and trying to blame Porton Down for everything,' Malcolm says. 'And now it seems you may be right.'

Silas raises his eyebrows at Strover. It's as close as he's ever heard Malcolm admit that he's wrong. 'How do you mean?'

'Caldicott did do a brief stint at Porton. Worked on human vaccines and therapies for serious pathogens. It all tends to be a bit niche over there – you either stay for life or get out quickly, worried that no one else is going to be interested in your narrow sphere of expertise. Only in his case, he was asked to leave.'

'Asked?' Silas says, surprised. 'What did he do?'

'Threatened to expose some experiments he deemed unethical. No legal protection for whistle-blowers in those days. They threw the whole jingbang book at him, said he could expect to spend the rest of his life in jail for multiple breaches of the Official Secrets Act. So he shut up and became a hospital pathologist.'

'Do you think he was right?' Silas asks. 'About the experiments being unethical?' He hasn't told Malcolm that Caldicott went on to have a second career as Jed Lando – with big pharma in America.

'I couldn't possibly comment – except to caution against judging events of the past by today's standards. Don't bother checking with Porton – they'll neither confirm nor deny he ever worked there. But I've got it on good authority. Any progress with the other two victims?'

'Not yet. And you?'

Silas was hoping Malcolm would have completed a full toxicology report by now.

'We still can't identify those substances found in their bloodstream – drawing a complete blank. I might send them over to Porton, just to see if they ring any bells.'

88

Bella

Bella pulls over in a lay-by, heading towards Southampton in the dark. She's on her way to Oxford to confront Dr Haslam about Erin, using the same satnav destination that her mum entered to pick her up from college at the end of term. But she's also in possession of sensitive material that she promised Jim she would get published. The story's far from complete – she doesn't have any proof that Erin was a victim of experiments at Harwell, nor does she know who made the crop circles – but it needs to be told before anyone else dies.

She puts the battery in her phone, dials the newspaper switchboard and asks to speak to Mark, her editor.

'He's off today,' a male voice says. Bella recognises him as Mark's frosty deputy, a man who barely acknowledges Bella's presence in the office, unless it's to put in his coffee order. 'Can I help?'

'It's Bella,' she says. 'I'm doing work experience on the lifestyle desk.'

The deputy says nothing. In the awkward silence that

follows, she hears him talking to a colleague, approving a page layout.

'I'm writing one of the "overheard" columns, down at a pub in Wiltshire,' she continues. 'The Slaughtered Lamb.'

More talking to someone else before he speaks to Bella.

'Nothing about that on the schedule,' he says, in an offhand way. 'As far as I can see, we've got the next two months of "Overheards" already sorted. You sure Mark commissioned it? We don't normally get temps to file copy.'

Bella has a terrible sinking feeling. Was Mark humouring her by sending her to Wiltshire? She closes her eyes, hears her mum's words on the phone to Dr Haslam: *She's just working as a secretary...*

'I wasn't actually ringing about that,' Bella says, pulling herself together. 'I've got a story for the news desk and was hoping Mark could tell me the best person to contact over there.'

'The news desk?' the deputy says, his voice dripping with derision.

'It's about the crop circle killings,' Bella continues, determined not to be beaten down by this condescending man. She's a journalist, not a temp.

'Bella, we don't know each other,' he says, 'but can I offer a word of advice?'

This time it's Bella who remains silent. His tone is beyond patronising.

'I know you're a family friend of Mark's, and went to Oxbridge and all that, probably a private school too, but the world's changing,' the deputy continues. 'Even the archaic world of newspapers. You've got to be good to get on these days, not just have a daddy with friends in high places. As

THE MAN ON HACKPEN HILL

for the news desk, it's shortly to go off-stone with the first edition, as any proper journalist would know. They might be a little busy right now.'

Bella feels a surge of anger. She's about to hang up but she can't let him get away with it. Not after mentioning her dad.

'First of all, my dad's dead,' she begins. 'He was killed in Mogadishu covering the Somalia civil war for *The Washington Post*, among others. Look him up – he was quite a good journalist in his day. And he despised nepotism of any kind. Secondly, I went to Clapton Girls' Academy – a secondary comp in north London – before going to an Oxford college with more than 90 per cent state-school entry, where I founded and edited a student magazine. Now put me through to the fucking news desk.'

89

Silas

'I can't believe Jed Lando used to work at Porton Down,' Silas says, turning to Strover. They are five minutes away from Cranham Hall, the psychiatric hospital where they suspect Lando was based.

'The boss won't be happy,' Strover says.

Silas is beyond caring about Ward. Malcolm's revelation has changed everything, refocusing their investigation on the secretive government facility. If Porton Down is connected with the crop circle deaths, so be it. Silas is more interested in Lando's history of whistle-blowing.

'What exactly did Malcolm say Lando did at Porton?' he asks.

'Human vaccines and therapies for serious pathogens,' Strover says, demonstrating an enviable ability to remember everything. 'Bubonic plague, anthrax, Ebola. If he was there now, he'd be working on Covid-19.'

But he's not. He's dead and hasn't worked there for thirty years. Silas turns off the main road and takes a narrow lane down towards Cranham Hall. Maybe

Lando still had military connections and was determined to expose ongoing malpractices at Porton Down. They could have been brought to his attention by Jim, who was convinced that the crop circles represent chemical warfare agents. And if, like many other pharmaceutical companies, AP Brigham has commercial connections with Porton Down, it would be in the interests of its smiling American CEO to stop Lando from jeopardising business.

'Any word from your boffins?' Silas asks, frustrated that they've yet to break the coded messages.

She checks her phone, shaking her head.

'A definitive answer would help,' Silas says, failing to disguise his impatience. 'The first one means this, the second means that. We'd know where we stand then.'

'They're proving much harder to crack than anyone thought,' Strover says.

Understatement of the year. The chemistry and mathematics professors have gone to ground, buried their eggheads in the sand.

'Jim says he knew straight away when he saw the circles,' Silas continues, angry now. 'BZ, LSD, VX. Boom, boom, boom.'

'And maybe he's right,' Strover says, 'if Lando once worked at Porton Down.'

Strover's phone vibrates with a text message. Silas taps the steering wheel while she reads it.

'It's from my mate in Hackney,' she says. 'A neighbour's just come forward, saw Bella's mother being "escorted to an unfamiliar car" outside her home earlier. Not sure how willing she was to get in.'

Silas shakes his head. Strover's like a dog with a bone, won't let Bella go. 'Still no word from Bella?' he asks.

'Phone's switched off.'

'Which college did she go to?' Silas asks. 'That's what they have at Oxford, isn't it? Colleges.'

Ward's always going on about his old Oxford college – he went there for a twenty-fifth reunion the other day.

'I didn't check,' Strover says. 'I just googled a local newspaper story about how she'd won a place at Oxford from her state school in Hackney.' She pauses. 'You didn't seem too convinced about her involvement in all this, so I left it at that.'

Strover is showing admirable restraint. Silas worked for a boss once who never took his theories seriously. Until he realised that was the point: incensed, Silas used to come back with more and more evidence until his boss had no option but to investigate.

'I'm still not convinced,' Silas says. 'All we know is that Bella has a friend called Erin, which is the name of the second victim, according to an anonymous caller to Crimestoppers. And she happened to meet Jim for a drink in the pub. It's not the most compelling evidence, is it?'

'And her mum seems to have disappeared,' Strover says. 'And she was once arrested on Studland beach, where Jim has just been taken away in an illegal ambulance.'

'Circumstantial,' Silas says.

Strover falls silent, checks her phone. Maybe he's pushing her too hard.

'Someone else has rung into Crimestoppers,' she says, reading an email. 'A tattoo artist in London. Read a tabloid story about the crop circle killings and remembers doing a load of rook feathers for a young Irish woman a few years back. No name, hers or the woman's, but apparently she had mental health issues.'

But the artist still went ahead and covered Erin's body in feathers. One of the many reasons Silas dislikes tattoos, the way they legitimise spur-of-the-moment decisions. Mel says he's being narrow-minded.

'OK, so if – and it's a big if – the second victim is Bella's friend Erin,' he says, 'and if Erin had "issues", then it's conceivable, as you suggested in the meeting earlier...'

'...that she might have passed through one of the psychiatric units run by AP Brigham,' Strover says, finishing Silas's sentence.

'Or been a volunteer for unethical military experiments at Porton Down,' Silas adds.

Silas is not sure where either theory leaves Bella. Did she know Erin's life was in danger and that's why she was in Wiltshire?

A siren starts up behind them. Silas glances in the rearview mirror and sees a modern private ambulance flashing its headlights.

'Alright, alright,' he says, pulling over to let the ambulance pass. 'Take a note of those plates, will you?' he adds, as the vehicle accelerates away. 'Check they're legal for blues and twos.'

He pauses, a thought taking seed.

'And run them against the ambulance seen on the beach at Studland.'

90

Bella

'Bella, can you hold for a sec?'

'Sure,' Bella says, glancing around the lay-by. She's been put through to the news editor on her paper but it's the car parked up ahead that's making her anxious. She exhales slowly, tries to calm down.

'Sorry about that,' the news editor says, coming back on the line. He sounds friendly and interested, everything that Mark's cold deputy wasn't. 'So tell me. What have you got for us? Mark's a good mate of mine – gave me my first break in journalism.'

And so Bella tells him the whole story, beginning with her Oxford college friend Erin, and how she's one of the unnamed victims of the crop circle murders, and ending with Jim, a government scientist, who believes that the coded patterns represent chemical weapons that are stored at Porton Down – and are still being tested on human guinea pigs at a secret facility in Harwell.

'Erin must have been experimented on – it's the only explanation,' Bella adds, worried by the growing silence at the other end of the phone. 'Everyone thinks that these sort of tests at Porton stopped in the 1960s, but they didn't. They're still happening today. And innocent people like my friend Erin are dying. I've got the whole inside story, from a government scientist, in his own words. And the tests are taking place in Oxfordshire, not down in Wiltshire. It's clever, very clever. Everyone's looking the wrong way. You still there?'

'I'm here,' the news editor says, his voice more reflective than before, almost sombre. 'Where are you now, Bella? Where are you calling from?'

Bella looks around her. The car in front hasn't moved. 'In a lay-by. On the A34. I'm heading to Oxford now. Once I've spoken to Dr Haslam, I can file my story, including right of reply quotes from—'

'Are you with anyone?' the news editor interrupts.

'No, why?' Bella checks behind her this time. There's no room for any more cars in the lay-by, thank God.

'That's quite some story you've told me,' he says, pausing. 'You might... need some help.'

'I'd rather not share a byline, if that's OK.' It's the last thing Bella wants on such a big story. 'I can always take it elsewhere, to another paper.'

Bella watches as a man gets out of the car in front and walks towards her. 'I've got to go,' she says.

'Is this a good number to ring you back on?' he asks.

Bella doesn't recognise the man but she has a bad feeling about him and puts one hand on the ignition key.

'This number's best,' Bella says. 'But I don't always have my phone on.' She pauses, eyes still fixed on the man. 'No one's managed to identify any of the crop circle victims. I knew one of them personally – my friend Erin. We're talking a front-page exclusive here.'

Bella presses her lips together. She's doing this for Erin. And for Jim too, wherever he is.

'I'm going to speak with Mark at home and then call you straight back, OK?' the news editor says, his voice full of encouragement. 'Thank you for thinking of us – and, you know, look after yourself, Bella. It's a little busy here right now but we'll be in touch shortly.'

She can't take a risk with the man. He's caught her eye and is approaching the car with purpose now, striding across the final few yards of the lay-by. Bella keys the ignition and jams down her foot, swerving past the man as she accelerates onto the main road in a plume of dust.

Five miles on, she pulls into another lay-by and retrieves her laptop from the boot. Pushing her seat back, she creates a new document and starts to type. She's not going to let the newspaper bring in someone else to write the article. *You might need some help.* Like hell she does. It's her story and she needs to tell it.

The words flow easily and after thirty minutes she's banged out the bare bones of a piece. Once she's checked it through and made some edits, she emails the copy to the news editor, copying in her boss, Mark, and adding a note that she'll file quotes from Harwell as soon as she's spoken to Dr Haslam and heard his point of view.

She still can't fathom her old tutor's role, but she's convinced that he's the missing link between Erin, the crop circles and the Harwell experiments. And whatever happens to her now, at least the media have Erin's name, know the identity of one of the victims. Jim's role will also be widely known, his courage acknowledged for breaking rank and revealing the government's ongoing programme of chemical weapons testing.

She puts her laptop on the passenger seat, takes a deep breath and sets off for Oxford, checking her rearview mirror for tails. It will feel strange to be back at her old college. Frightening too. But she owes it to Erin.

91

Jim

Jim stares at the ambulance doors, waiting for them to open, barely able to breathe. They have come to a halt after several hours on the road. He hears the driver jump out and walk around to the back of the vehicle. Jim glances at the two men sitting either side of him, his heart thumping so hard it hurts, but both remain seated.

'Are we getting out?' Jim asks, trying in vain to move his arms in the straitjacket.

'First you have to decide if you're coming quietly,' the man with the goatee says, unbuckling himself.

'If you tell me why I've been brought here,' Jim says, short of breath. Has MI5 really driven him back to Harwell?

Both men stand up, folding their seats away like commuters arriving at their destination.

'There's been some controversy about the work you were involved with here,' the man says. 'You need to keep your head down for a while – for your own safety.'

The other man turns to the filing cabinet and starts to peel off the tape holding the syringe. A loud tearing sound reverberates around the ambulance's metallic interior. God, Jim hates syringes. Holding it up to the light, the man checks the level and releases a small amount of liquid. He must know he's being watched, must know how injections haunt Jim day and night. If it wasn't for the tightness of the straitjacket, his whole body would be shaking like a jelly.

'Controversy?' Jim says, managing a dry, frantic laugh as he looks away from the syringe. 'You don't understand, I've already given the story to Bella. I've told her everything about Harwell. It's too late for keeping bloody heads down. Bella works for a national newspaper. It will be on the news tomorrow – "Exposed: secret chemical weapons experiments on human guinea pigs at Harwell." The cat's already out the bag.'

He's gabbling, speaking way too fast, but the thought of Bella gives him strength. She's out there somewhere, sharing his crusade, fighting dark forces for justice. The two men stand by the doors, waiting for them to be opened.

'Bella's a bit of a fantasist, I'm afraid,' the man says. 'She's on work experience for the paper. As a secretary. I don't think we'll be reading any front-page stories by her just yet.'

'She's good,' Jim says, ignoring the man's smirk, praying that Bella will have read through all the files on his USB by now. 'You won't be able to stop her. And she knows the identity of one of the victims.'

Both men glance at each other. That's got their attention. Was it a mistake to mention it? They will go after Bella harder now, if they haven't found her already.

'They were at college together,' Jim adds, the damage already done. His arms start to shake uncontrollably inside the straitjacket.

A click and the double doors swing open on the evening. Dusk has fallen since he left Studland but the buildings in front of him are illuminated with spotlights, including the blurred but unmistakable profile of the modern laboratory block at Harwell, home to the biosafety level four high-containment facility. Jim shakes his head in despair. It's true. He's back as a 'volunteer' – as punishment for leaking classified information to Bella. MI5 must have a dark sense of humour, silencing him with the very experiments that he was trying to expose.

'Home sweet home,' the man says. 'These might help.' He slides Jim's glasses onto his nose, pushes them into position and stands back to admire his handiwork, as if he's an artist putting the final touches to a sculpture.

Harwell comes into focus, the buildings sending a bolt of fear through Jim. They are at the rear of the high-containment facility, where the supplies are brought in. His final days here were good but he wishes he could forget what came before. For a while he couldn't remember, but moments continue to return, isolated vignettes of terror. He should keep writing them all down but Bella's got more than enough material if she's managed to open his Harwell diary.

'How does it feel to be back?' the man asks, unlocking Jim's wheelchair from the floor.

'Just wonderful,' Jim says.

'And have we agreed to come quietly? No repeats of what happened on the beach?'

Jim glances again at the needle. The other man is holding it across his chest like a pistol. The last thing he wants is to be drugged by whatever's in it. BZ, VX, maybe even LSD.

He needs to stay focused, keep his wits about him until the story's out and he's been released, hailed a hero, a true public servant. One day, in a lucid moment, his dad will understand what he's done here, why his son had to break ranks. He will be proud of him. And they will hunt prime numbers together again, and race to finish Rubik's cubes.

Jim nods at the man, swallowing hard as he is wheeled down the ramp into the night.

'Welcome back,' a familiar voice says, sending Jim's body into another spasm of fear.

Dr Haslam.

92

Silas

Silas pulls up in the visitors' car park outside Cranham Hall and looks over towards the main building, lit up against the evening sky. It must have been a country house once, complete with a sweeping drive leading up to a gatehouse, beyond which is a gravel courtyard. The windows of the house appear soulless, dark and foreboding. A high and illuminated brick wall runs around the boundaries of the estate – once for keeping intruders out, now for keeping patients in.

'I prefer somewhere a little cosier myself,' Silas says to Strover, sitting next to him.

'Built circa 1850 in the Tudor Gothic style, with a Jacobean themed staircase, it was home to the first Duke of Temple Grove,' she says, reading from her phone, 'then an all-boys prep school, a drugs rehab centre and finally a psychiatric hospital in 1992.'

'Catering to the aristocracy's every need,' Silas says. 'Let's talk to security and take a look round.'

Five minutes later, Silas and Strover walk across the

courtyard and enter the main reception area, where a woman behind a desk takes their names and asks them to wait for the duty manager, who will be down in a minute. No patients are in sight and the place is eerily quiet, smelling more of old furniture polish than hospitals. Silas doesn't expect the manager to say much, suspecting that the grinning fool of a CEO didn't buy into their trip to Newcastle and has already put in a call to warn staff of their arrival.

'I'm sorry to delay you.'

Silas looks up to see a woman walking down the formal staircase. She strikes Silas as an unemotional, disciplined manager, hair tied in a neat bun, blue medical uniform crisp and clean. Which makes it more of a surprise when he notices that she's been crying.

'DI Hart, DC Strover, Swindon CID,' Silas says, turning to his colleague. 'We've come about Jed Lando.'

'I know, I've just been told,' the manager says, casting her eyes down at the wooden herringbone floor. The CEO didn't hang about.

'I'm sorry,' Silas says. 'You knew him well?'

She nods, wiping at her nose with a tissue that she tucks away in her sleeve.

'What was the nature of his work here?' Strover asks, stepping in with a more sympathetic tone. Silas isn't good with tears.

'He was our boss,' she says, turning away. Her accent is Scottish, maybe Borders. 'Until last week, anyway.'

'Before he left the company?' Silas asks.

The woman nods. 'He was on holiday.'

Silas can sense what she's thinking. Shabby to fire some-one when they're on holiday.

'And he had an office here?' Strover asks.

She nods again. 'Upstairs.'

'Mind if we take a look?' Silas asks.

His tone makes it clear it's not a question. He glances out of the window onto lawned gardens as they follow the manager up the Jacobean staircase to Lando's office. The CEO must have told her to be accommodating to the police.

'Are you full at the moment?' he asks, making a poor attempt at small talk. It sounds like he's asking about a bloody hotel.

'We're always full,' she says.

'Just that we haven't seen any patients around.'

'The women's wing is in lockdown right now – everyone's in their rooms. We've a separate ward as well, a high-dependency unit. They're more secure.'

Silas doesn't ask the reason for the lockdown. It could be anything from a blown fuse to an escaped patient, if Conor's experience of secure units is anything to go by. He's in a very different facility in Swindon, less secure but somehow more medical. At least it felt that way under the old management. Purpose-built, modern. This place doesn't feel right, too many echoes of a Victorian asylum.

'But patients can wander around, inside and outside?' he asks.

'Some can,' she replies, smiling thinly. 'It depends how well they are.'

Jed Lando's office offers no clues, nothing obvious to suggest his previous career as a pathologist. Nothing to point towards a life beyond work. No family photos or handwritten letters on the wall from young children, like Ward has from his godchildren. Some unopened mail and a couple of invoices on his desk, along with a colourful patchwork of Post-it notes with phone messages that will never be answered. He hadn't cleared his office because he thought he was coming back. The only curiosity is an old Remington typewriter on a filing cabinet in the corner.

'Do you have mortuary facilities on site here?' Silas asks, trying to make light of the question.

'A mortuary?' the manager asks, surprised. 'There's a cold room in the basement but—'

'Who has access to it?' Silas interrupts.

'The duty manager, senior nurse. It's very rare that we need it.'

'Did Jed Lando have access?'

'I don't know.'

'But you do?' Silas asks, making it clear that once again they will need to take a look.

The manager hesitates for a second and then heads for the staircase. Silas is about to follow her out of the office when a small photo, cut from a newspaper, catches his eye on the wall above Lando's desk. It's of a bearded man and there's a caption below it: 'Stefan P. Kruszewski, clinical and forensic psychiatrist.' Silas peels it off the wall and follows her down the narrow spiral steps to the basement.

It's hard to imagine a body being manoeuvred around the staircase's tight bends. Almost impossible. He glances at Strover as the manager stops at the end of the dark corridor and juggles a bunch of keys. Is this the lock that the key fits? Selecting the right one, she opens a heavy door and turns on a light. The room is a fraction of the size of the Great Western mortuary and much cooler, like a cold store. Maybe it was used by the house's kitchen staff before electric fridges. A place to store game pies and plucked pheasants. Silas feels hungry at the mere thought. A stainless steel table runs along one side and various tall cupboards and chairs have been stacked up against the far wall.

'We use it as a storeroom, as you can see,' the manager says.

'When did the last person die at Cranham Hall?' Silas asks, taking in the room. Strover shoots him a look. 'The last patient,' Silas adds.

'Three, maybe four years ago,' the manager says. 'Just before I joined. The place was under different management.'

'What's through here?' Silas asks, looking behind a large wardrobe at an old door in the brick wall – and at its lock in particular.

'Through where?' the manager asks, coming over to Silas. She'd make a lousy actor. He nods at the door. Hard to miss but in fairness it looks as if it hasn't been opened for a while.

'I wasn't aware of a door here,' she says. 'It must lead outside.'

Silas glances across at Strover. There's no need to say anything. She reaches into her pocket and holds out a small, clear plastic bag with the old key in it. Silas slips on a pair of purple forensic gloves, removes the key and slides it into the keyhole.

93

Jim

Jim recognises several of his fellow scientists as he's wheeled into the modern, whitewashed common room where he used to go for a break from the test chambers. His former colleagues all look in bad shape, worked to the bone. Jim smiles at a man he used to know quite well, but he stares back at him, his eyes two pools of emptiness. Soon Jim will be at their mercy, a volunteer for whatever experiments they have in line for him.

Jim's agreed to come quietly, fearing the needle, but he has a sudden urge to address these people, to tell them the game's up and the world will soon know what's happening here. The two MI5 officers in the ambulance assured Jim that he would be treated well if he cooperated, but he knows his fate. His plan is to wait until the time is right, and then tell his former colleagues about Bella and the information he's given her. They won't like it, the loyal ones, but others will be relieved that this hell might finally be over.

'Where are you taking me?' Jim asks, as he leaves the common room behind and is pushed down a long, familiar corridor. The high-containment facility's walls are sterile and white, punctuated with metal doors on either side, reinforced glass panels allowing a view of the test chambers inside.

The man pushing him leans down to his ear. 'To your old room,' he says.

Jim's arm muscles start to spasm beneath the straitjacket, pressing against the canvas ties. It's not a room. It's a test chamber, the place where he worked, day and night, week after week, conducting experiments on his volunteer. Only this time, others will be experimenting on him.

'I need to see a lawyer,' Jim says, rocking from side to side in his wheelchair. 'I demand to see a lawyer.'

He would have seen one as a matter of course if MI5 had taken him directly to a police cell. But here, hidden away at Harwell, the law suddenly feels very distant.

'You'll get the needle if you don't shut up,' the man says, waiting as the door to his chamber is unlocked and he is wheeled in. Not much has changed. A simple mattress, whitewashed walls. The only new feature is a mirror and basin.

Two well-built men walk into the chamber behind him and together they unstrap Jim from the wheelchair and extricate him from the straitjacket.

'Thank you,' Jim says, but he knows they haven't freed him out of kindness. He shakes the numbness from his arms, rubbing where the straitjacket's thick canvas has chafed his elbows. There's no point trying to make a run for it. These men are too strong, his own body too heavy with adrenaline. And they are already tying his hands together with a leather strap and fastening them with a tether to the bed, before they file out of the chamber and lock the door in silence. Jim looks over to the observation window, wondering who will soon be watching him, what they will do to his mind.

It doesn't take long to find out. Moments later, the door is unlocked again and a big bulk of a man enters, whistling a delicate tune. He's wearing blue overalls and has a floor mop in one hand and a huge grin on his face. Everything about him is oversized: his bony frame, sloping forehead, dark, hound-dog eyes. Jim recognises him at once as Vincent.

'Yo! Our resident scientist,' Vincent says, revealing a mouth full of oversized teeth and a south London accent. 'I heard you was back. Nice to see you again, Jim-boy. Meant to clean this room before you arrived but we've had a lockdown in the old wing and I'm running late.'

Jim doesn't fall for the room-cleaning routine – or the pally tone. He's seen it before, used a similar approach himself when he was trying to surprise his own volunteer.

'I don't need my room cleaned,' Jim says, shifting to the edge of the hard bed. The tether from his hand strap to the bed frame is about two metres, allowing him to stand up and reach the middle of the room.

'Fair enough,' Vincent says, sniffing the air. There's a hint of Italian in his voice too. 'A little bit ripe in here, isn't it? Not sure why they put you in this room, to be honest. Might just give the place a quick freshen up.'

Jim's eyes widen as Vincent pulls out an aerosol spray from the leg pocket of his overalls.

'Stop!' he shouts, lunging forward to try to grab the can from the man. His tied hands make him more clumsy than usual.

'Easy now, Jim-boy,' Vincent says, holding the can above Jim's outstretched arms as if he's playing with a dog on a lead. Jim's tall but Vincent is even taller. His cheery smile has gone, replaced by a taunting smirk.

'Do not spray that in here,' Jim orders, breathless, still trying to reach up to the can, his hands held back by the harness.

'Security,' Vincent shouts casually, as if they've entered the next stage of a game. The two men begin to wrestle, Jim still trying to take the can off him. Vincent hits a button on the wall and an alarm starts up. 'Security!' he repeats, calling out with more urgency now.

The sound of running feet in the corridor. And then the room is full of men, who throw Jim onto the floor and pin him down.

'You didn't tell us it was our old friend Jim,' one of the men jokes. Jim recognises the voice – the same man who had held his face down in the mud beside the bench, outside the high-containment facility. 'Turn that bloody attack alarm off will you, Vince.'

'Let me go,' Jim shouts.

'You're not going anywhere,' the man says.

'I was only trying to freshen up his room,' Vincent says.

'Let me go,' Jim shouts. His left leg is being bent out sideways at an increasingly awkward angle. 'Let me fucking go.'

But it's too late. He hears his knee socket pop and screams out in agony.

'Alright, that's it,' the man says. 'Give Jim a shot.'

'Can I spray the room now?' Vincent asks.

'Shut it, Vince.'

Jim squints up from the floor, where his cheek is pressed against the ground. His glasses have been crushed again, fragments of lens cutting into his cheekbone. Why do they always break his glasses? Someone has started to grind their knee into the base of his skull.

'You're hurting my neck,' he cries out. His own knee is also on fire with pain.

It's probably BZ in the aerosol can. Maybe the Boomer. At Edgewood Arsenal in 1964, the Americans considered taking out an entire trawler ship with aerosolised BZ. Code name? Project Dork. All Jim knows for certain is that a needle has pierced the skin of his thigh and the world is starting to fade.

94

Silas

The key turns easily in the lock and the first thing Silas sees is the floor-to-ceiling refrigerated cabinet in the corner of the hidden room.

Strover comes through the door behind him, followed by the manager.

'Do you have a warrant?' she asks.

'Anyone in the chiller?' Silas replies, ignoring her. 'Seems like Cranham Hall does have its own mortuary facilities after all.'

'I need to speak to... to someone in authority,' the manager says.

'Like your American CEO?' Silas offers.

'I had no idea about this room,' she continues. 'And you have no right to be in here without a search warrant. It's a place of rest, for heaven's sake.'

'Except that apparently no one's been at rest for four years. I'll come back with a warrant, don't worry. But before I leave, we need to see who's in there.'

He nods at the cabinet in the corner. Warrant or no

warrant, he's allowed to enter and search any premises to 'save life and limb'.

'No one's died here, detective,' the manager says, as Silas walks over to the unit.

'In that case, there's nothing to hide.'

He takes hold of the chrome handle, sleek and long like an American fridge's.

'I must insist that you show some respect,' the manager says.

'Jed Lando was murdered,' Silas says, turning to face her. 'He's a boss I sense you were fond of. We're trying to find his killers. Let's show him a little respect too, shall we?'

Silas opens the heavy door, releasing the rubber seal with a sucking hiss. A blast of cold air hits him, followed by the faint smell of bleach. For the second time in twenty-four hours, he finds himself looking at a stack of five body trays. Only this time there aren't any corpses. Maybe the manager was right about no recent mortalities at Cranham Hall.

Silas stares at the trays. They might not always have been empty. Is this where the victims were kept before being taken by Lando to the crop circles? It won't be straightforward for forensics to establish. The cabinet is for the long-term storage of cadavers – minus 50 degrees, according to a display screen – but it's the smell that worries him. Bleach is one of the easiest ways to degrade DNA.

Back outside, Silas and Strover stride across the floodlit courtyard, watched by the manager. He's beginning to doubt whether she did know of the hidden mortuary, even though it has outside access, double doors leading onto a small service courtyard.

'Were you expecting to find other bodies?' Strover asks, struggling to keep up with Silas.

'Not if they knew we were coming,' Silas says. 'They could have been moved elsewhere between our visit to Reading and arriving here – it certainly smelt like it. Get on to TVP and ask for uniform backup. I'll call the boss and sort a warrant. We need this whole place searched from top to bottom.'

TVP – Thames Valley Police – won't be happy that their Wiltshire colleagues are on their patch. Ward won't be too pleased either.

'How about patient records?' Strover asks. 'We could see if anyone called Erin stayed here.'

'If Erin died at Cranham Hall, they might not even have recorded it,' Silas says. 'It's one of the great scandals of our time – the way deaths in places like this are investigated. Or not.'

Silas was once asked to look into a suspicious fatality at a psychiatric hospital, and was shocked by the lack of an independent investigation at the pre-inquest. The coroner had to rely on evidence gathered by the very hospital that was under investigation. And that was a death the coroner had actually got to hear about. If the deceased were homeless, had no family or friends, as seems to be the case with the two crop circle victims, their passing might go unrecorded, their lives never missed.

'But why lay their bodies in a crop circle in Wiltshire?' Strover asks. 'Surrounded by mathematical codes that no one can decipher?'

Before Silas can answer, a cry of agony rings out in the darkness from behind the main building.

'Jesus, someone's not happy,' Silas says.

A second cry, louder this time. They sign out at the gatehouse and hurry around the outside of the perimeter wall, watched by a member of the security staff.

'That sounded disturbingly familiar,' Silas says, thinking back to the shout they'd heard outside the Slaughtered Lamb. Not the cry itself, but its tone, the timbre of pain. They reach the back of the main building, where a big, modern extension has been constructed adjacent to the old Victorian house. It looks more like a hospital, purpose-built and floodlit. Silas thinks of Conor again as he tries to see over the wall. And then two gates swing open and a private ambulance sweeps out.

'Tell me that's not the number plate of the ambulance at Studland – the one that overtook us on our way here.'

'I'll chase up Dorset now,' Strover says, pulling out her phone.

'And?' he asks expectantly, when she hangs up.

'It's the same,' she says.

'Oh Christ,' Silas says, nodding at the modern extension. 'What the hell's Jim Matthews doing here?'

95

Jim

'What were you thinking? I heard his screams from my office. You are aware we have two police officers on the premises.'

'I'm sorry—'

The unmistakable slap of someone being hit hard across the face.

Where are these voices coming from? Inside Jim's head? He's heard enough of them in his time.

'Don't "sorry" me. Now fuck off, the lot of you.'

Haslam. Dr Haslam. One moment so meek and mild, almost fey with his long hair and round glasses and moist little eyes, the next, he's smacking you in the mouth and shouting like a demented sergeant major. Christ, he hit Jim enough times when he wasn't happy with his progress. Jim suddenly feels vulnerable as he hears the footsteps fade. He's lying on a bed, unable to move, with just Dr Haslam for company. And his left knee is still burning up with pain. Is that what woke him?

A phone starts to ring. Jim opens his eyes, but all he

can see is a blur, edged with crimson. He feels slow, heavy-lidded, as if he's drunk gallons of treacle and it's filled every limb of his body, pulling him down through the bed. The phone's still ringing. Is it his phone? Ringing in his head?

'Haslam speaking.' A long pause. 'I know they're bloody on site, we've been tracking them... In the old wing.'

Jim's head is hurting too. He manages to lift a hand – so heavy – to touch his forehead. No fresh blood.

'I told her not to show them anything... It's locked, don't worry. And everyone's been moved. I'm also the only one with a key.'

If Jim strains, he can catch the voice at the other end of the phone. No words, just the accent. American. Maybe Haslam *is* in his head. The American too. They used to talk like this, the hostile voices that argued day and night, telling him what to do, passing judgement on his life. His persecutors, his only companions.

Jim tries again to work out what's happening. Haslam must be somewhere outside, in the corridor. The test chambers are all soundproofed, which means Jim's door is open. It explains the faint draught on his face. He tries his eyelids again, but they don't respond. Shut like a pair of heavy garage doors. Why can't he bloody see? Even without his glasses, he can usually make out something, dim shapes, light and darkness.

Haslam is talking again, his voice agitated.

'We just need to ring-fence Lando,' he says. 'He had a breakdown, commissioned some weirdo, hippy-shit circles in a wheat field and did some shocking things with two dead bodies before tragically dying himself in hospital... We knew nothing of what he was doing. Which is true. We were as surprised and as horrified as everyone else. And there's no evidence to link the victims to Cranham Hall. Identities unknown – no one knows who they are, the police, the papers. Jed Lando knew but he can't talk any more... I don't know if she really knows. I'm trying to find out.'

Jim's eyes are opening. He blinks and blinks, forcing his retinas to work as he turns his head. Slowly, the blurred edges of his chamber start to sharpen, not by much but enough. The open door, the empty chair beside him, the diminutive silhouette of Dr Haslam on his phone in the corridor.

'I'll talk to the two detectives if I have to... No, just Jim Matthews. We're expecting the woman any minute. When she's in, we'll stay in full lockdown... I'm confident the situation can be contained once we have them both under the same roof... Of course I'm not going anywhere.'

Dr Haslam comes off the phone and enters the room, pulling up the chair to sit in front of Jim. He's still a blur but Jim knows it's him because of the sickly sweet aftershave. Dr Haslam never goes anywhere without it. Maybe it's to mask the smell of other people's fear.

96

Silas

'When did CSI say they'll be here?' Silas asks Strover, as they return to the car park at the front of Cranham Hall.

'Twenty minutes,' Strover says, glancing at her watch.

'You did well,' Silas says. Even though CSI only has to drive over from the Forensic Investigation Unit in Oxford, it's a quick response for so late in the day.

'Everyone wants a slice of this case, guv,' Strover says.

Did she just call him guv? Finally? They are both buzzing, sensing a major breakthrough, now that they know the key found on Jed Lando's body fits the lock of a mortuary in the basement of Cranham Hall.

Silas looks back at the old building again as they get into their car. Once he's been issued with the warrant, he'll search the place for Jim Matthews.

'Jim's somehow central to all this, I'm sure of it,' he says.

'Aren't you forgetting Bella?' Strover says.

It's true. Silas is still to be convinced of the journalist's relevance to the case. He's about to apologise when Strover's phone rings. Her face lights up as she switches the call to speakerphone.

'I'm with my boss, would you mind repeating that?' Strover says. And then she whispers to Silas: 'The chemistry prof.'

'We've finally had a breakthrough,' the professor says, as Silas leans over to listen. The professor sounds positively animated, a far cry from the world weariness of their last exchange. 'We've continued to reverse-engineer the encoded messages, working on the assumption that it was either the molecular formula for a known chemical warfare agent or a prescribed antipsychotic.'

'And?' Strover says.

Silas taps the steering wheel, unable to cope with the tension.

'All three patterns relate to what we believe are antipsychotic medicines.'

'Are you sure?' Silas asks, unable to resist joining the conversation.

'We're sure. There's just one catch: each one is a variant of a readily available second generation neuroleptic, altered at the molecular level to create something that we don't believe is yet on the market. We're not talking big variations, but it appears to be about dropping the percentage of blockaded $D2$ dopamine receptors to below 60 per cent. The goal – and I'm speculating here – would be to eliminate all risk of negative side effects, while allowing for a reduced, possibly selective, treatment of a patient's more severe psychotic positive symptoms. Dangerous delusions, for example.'

Silas closes his eyes. He knows all about dopamine. When Conor was diagnosed with schizophrenia a few years back, he and Mel had endless meetings with psychiatrists as they tried to get the right balance of medication for him. Hit the sweet spot. Block 60 per cent of the dopamine receptors and the hallucinations and delusions start to decrease, but once you get up to an 80 per cent blockade, 'extrapyramidal' side effects kick in, including muscular spasms, restlessness and tremors.

'OK, so no brand names,' Strover says to the professor.

'These are essentially new drugs,' the professor says. 'Untried and untested and potentially lethal as a result. Unless they're the subject of ongoing clinical trials somewhere, but I've banged the jungle drums and nobody's ever heard of them. They usually take years to reach the market. And something like this would be the holy grail of antipsychotic medication – to cure psychosis without any side effects. But, as far as I understand, you can't really pick and choose which delusions remain and which ones are eliminated.'

'Thank you,' Strover says. 'For all your help.'

Strover signs off and turns to Silas. Her team of boffins has finally come good.

'So not Porton Down,' she says, sitting back. 'At least the boss will be happy.'

'Google this for me, will you?' Silas asks, passing Strover the newspaper cutting he took from Lando's office. He doesn't give a damn about the boss.

'Stefan P. Kruszewski, ' she says, reading the caption as she types it into her phone. 'A good name to google... He's a whistle-blower in America. Something of a hero. Exposed malpractice at a psychiatric unit, as well as at two well-known pharmaceutical companies.'

It's the confirmation Silas was hoping for. 'Jed Lando's target wasn't Porton Down – it was big pharma,' he says, thinking of the dramatic crop circles again, the bodies placed inside them. He remembers when he took the call about the first victim, a young man, and thought it might be Conor. An image of him in a wheat field comes and goes. It could so easily have been his son. 'We need to ring Malcolm, ask him to check the victims' bodies for these three drugs.'

Silas pauses. Something else is troubling him. He thinks again of Jim, the young scientist's insistence on working at Porton Down; the secret base's connection with Lando and his crop circles; AP Brigham's apparent interest in him; the low dopamine blockade of untested antipsychotics; the high chance of delusions. And a terrible thought begins to take seed.

97

Jim

'So, how was... Porton Down?' Dr Haslam asks, sitting beside Jim's bed in the test chamber. 'The furry little white rabbits?'

'Why am I back here?' Jim asks, squinting at Dr Haslam, aware of his stomach tightening. His own voice sounds like someone else's, thin and distant. He wasn't just working with rabbits.

'Where do you think "here" is?' Dr Haslam asks.

'Harwell, of course,' Jim says, puzzled by the question.

Dr Haslam sits back, shaking his head in disbelief.

'Truly remarkable,' he says. 'You're back here – at Harwell indeed, why not – because someone was trying to cause you harm. Us too. And we couldn't have that.'

'The harm is already done,' Jim says. 'I've told her everything. About the experiments you're doing here. VX, BZ, LSD.'

Jim desperately wishes he could see Bella, hear how she's filed her story, share the sense of anticipation as the news presses start to roll.

'And who's "her"?' Dr Haslam asks, with almost panto-mime curiosity.

'Bella. She's a journalist.'

'Of course, dear Bella the journalist.' Dr Haslam glances at his watch. He's not taking Jim seriously. 'She should be with us any minute.'

With us? Jim tries to sit up in bed but the weight of his own body is too much, the pain in his knee too intense. Bella can't come here. She must be kept away from Dr Haslam.

'Why's she coming to Harwell?' Jim asks, his voice cracking in desperation. It's too dangerous. If her best friend Erin was experimented on, Haslam could do the same to her.

'I imagine she's a diligent journalist and wants to hear the other side of the story,' Dr Haslam says. 'Give us our "right to reply".'

Jim somehow needs to get a message to Bella, warn her to stay away.

'It was a cruel thing to do, putting you two in touch,' Dr Haslam continues. 'But he was obsessed with *folie à deux*, you see. How did he do it? Write you a letter?'

Who was obsessed with *folie à deux*? And how did he do what? Jim doesn't understand the questions. His brain is not functioning properly. It's like there's a time lag between synaptic connections, an echo on the neurological phone network. Dr Haslam's right, though. Someone must have put them in touch. It's the only explanation for their encounter in Wiltshire.

'She found me in a pub,' Jim says.

'That was lucky.'

'She must have been sent.'

Dr Haslam sits back and sighs. 'For once, I fear you're right.'

What does he mean, 'for once'?

'And are you going to tell me who?' he asks, his tone cold now, dismissive.

'I don't know who it was,' Jim says.

Dr Haslam says nothing. His silences were often the most frightening part of a session with him.

'What are you going to do with me?' Jim asks quietly.

Dr Haslam leans forward to peer into Jim's face. Jim screws up his eyes, recoiling from the warmth of Dr Haslam's breath on his cheek.

'Tut, tut, I see you've been banging your head again,' he says, touching Jim between the eyebrows. It's sore from his earlier wound and Jim winces, jolting his head back. 'I thought we'd put a stop to all that nonsense.' He unlocks the tether tied to Jim's hands. 'Get up.'

Jim's eyes open. His response to the familiar barked command is visceral, like a triggered reflex, and he feels his body already trying to move.

'I can't,' he says, his muscles straining in vain.

'Get up,' Dr Haslam repeats, deaf to Jim's protests.

'I can't move,' Jim says, but slowly, very slowly, muscle memory takes over. Jim turns onto his side and lifts himself up on one elbow. Knee throbbing, he slides his legs off the bed and onto the cold tiles. It's not his old chamber, which had wooden floorboards. He feels a hand on his arm as Dr Haslam helps him to stand.

'This isn't my room, is it?' Jim asks, swaying on his feet.

'We've upgraded you,' Dr Haslam says. 'Given you a few perks – a basin and mirror. Actually, it's for staff. You shouldn't be in here at all but we're full. And you're an important government scientist. At least, you were. Out there in the real world doing very well for yourself – until someone put you in touch with Bella.'

Jim can hear the rising anger in Dr Haslam's voice as he helps him to walk over to the basin. For a moment, the physical contact, one arm around his shoulder, feels almost friendly, but then Dr Haslam moves his hand up to the back of Jim's head and gathers a clump of hair in his fist.

'Take a good look and tell me who you see,' he says, pushing Jim's face close to the mirror.

Jim's legs start to buckle. Gripping the sides of the basin to steady himself, he peers into the mirror, close enough for his breath to create a small, pulsing cloud of condensation on the glass. A haunted face stares back at him. Dark eye sockets, a wounded forehead. He barely recognises himself.

'Who is it?' Dr Haslam repeats. 'Who do you see?'

Jim closes his eyes and opens them again. 'James Matthews,' he whispers, remembering the routine with a shiver. 'Important government scientist.'

'Louder,' Dr Haslam shouts, twisting Jim's hair in his fingers.

'James M—' But before he can repeat his name, Dr Haslam smacks Jim's forehead into the mirror, fracturing the glass into a jigsaw of shards.

'Now who do you see?' Dr Haslam asks.

Jim tries to lift a hand to his head, anything to stop the pain coursing through his skull, but he hasn't got the strength. Instead, he slumps forward over the sink, watching droplets of blood explode against the porcelain. Slowly, he looks up at the multiple images of himself staring back at him.

'James Matthews,' Jim says.

'The old James Matthews,' Dr Haslam corrects, still holding the back of his head. 'Psychiatric inpatient, diagnosed with paranoid schizophrenia. But I made you well, didn't I? No negative symptoms, and only a few positive ones, the benign delusions that we all experience at some time in our lives. I sent you back out into the world, even found you a job. So the least you can fucking do is tell me who put you and Bella in touch. How did she find you? Was it Jed Lando? Or someone else? Another traitor?'

Dr Haslam is shouting now, calling for Vincent to remove the remains of the mirror from the wall and clean up the mess, but his words are fading, and Jim is no longer at the basin. He is outside in the corridor, looking in through the observation window on a young volunteer as Dr Haslam smacks his fragile face into the broken mirror again and again, repeating a question that he can never answer.

98

Bella

Bella pulls into the college car park, the same corner slot that her mum had used when she came to pick her up at the end of term. It feels a long time ago now, before the summer, her shifts at the migrant centre, hanging out with her mum at home. Getting to know each other again. Bella was ready to leave uni and face the challenges of the outside world, to pursue a career in journalism. She never imagined that she'd be back at her old Oxford college investigating the death of her best friend.

She gets out of the car and walks over to the lodge, wondering if the porters will have forgotten her already, or whether they've been warned by Dr Haslam to deny her access. It's strange being back in college, a weird mix of dread and satisfaction. She knows that Dr Haslam was responsible for preparing her for real life, and there was a time when she was grateful to him. That's all changed now. She's angry, ready to confront him about Erin. About Harwell too. What the hell was he doing over there?

The night porter on duty looks up at her as she walks into

the lodge, his lips breaking into a faint smile of recognition. Bella had a love-hate relationship with the porters during her three years as an undergraduate. By the end, she had earned their grudging respect, the only way to get on with them. If you showed any weakness, they could make student life a misery, as she discovered in her first year.

'Bella,' the porter says. His tone is not exactly welcoming but nor is it aggressive. The other porter behind him looks up from his bank of TV screens. She doesn't recognise either of them.

'I've come to see Dr Haslam,' she says, wondering how the man knows her name. 'Is he around?'

'He's expecting you,' the porter says, pushing forward a clipboard for her to sign in.

'Expecting me?' Bella says, unable to conceal her surprise. At least he's back from Italy.

The porter glances at his watch. 'He's in the library.'

Bella signs in, trying to stay focused as she walks through the familiar glass sliding doors of the porter's lodge and across the floodlit courtyard. A CCTV camera mounted on a wall in the far corner tracks her progress. Student security became a big issue in her time here. She glances around the Victorian building and remembers a night when Erin was more off her face than usual. Round and round the courtyard she ran, arms outstretched like a bird, chased by three breathless porters.

'You can't catch me, you bunch of dryshites,' she'd cried. If it wasn't so sad, it would have been funny.

Bella's room was on the second floor and she instinctively glances up at the window. The light's on, which is disconcerting. Who's in there now? Maybe the rooms are being hired out for a summer school. She steps into the reception area, where a woman behind a desk looks up at her. She's unfamiliar too. Another CCTV camera clocks her arrival.

'I'm meeting Dr Haslam, in the library,' Bella says.

'Welcome back,' the woman says, smiling at Bella. 'He's waiting for you.'

Bella's eyes linger on the woman, testing the strength of her smile. Sure enough, it soon starts to collapse into a look of guilt and anxiety. The woman turns away.

'What's wrong?' Bella asks, fear rising.

'Excuse me?' the woman says, glancing up again.

'You're not telling me something,' Bella says. Her ears start to hum, a loud, persistent noise as if someone's just hung up on her.

'Dr Haslam's waiting for you,' the woman repeats, keeping her head down as she types.

Bella walks along the lit-up corridor, tracked by more cameras as she turns right into the library, still spooked by the woman's reaction. She came here a lot when she was an undergraduate, spent hours reading on the windowsill, the shadow of the thick glazing bars falling across her like a crucifix in the afternoon sun.

At first she thinks the library is empty. The lights are off, the rows of bookcases standing sentinel in the shadows. She looks down one aisle, and then another, before she sees him at the far end, leaning against a desk in a solitary pool of light. Dr Haslam, the same as ever, patches on his corduroy jacket, small round reading glasses. Erudite eyes at odds with his soft, pale skin. Something of the sorcerer about him. He doesn't look up. Instead, he stays focused on the book in his hand, and starts to read out aloud as Bella approaches.

'Her eyes are wild, her head is bare, The sun has burnt her coal-black hair.'

Bella's in no mood for poetry, not even Wordsworth.

'What happened to her?' she asks, looking around again. No one else is in the library. 'Tell me what you did to Erin.'

99

Jim

Jim hears Vincent before he sees him. Only Vincent whistles like a blackbird in spring. He opens his eyes, watching as a blur moves around in front of the basin on the other side of the chamber. He's still in the same place but Dr Haslam has gone, leaving him with a nauseous headache. Slowly raising his tied hands to his forehead, he touches the soft material of a wet bandage, wrapped around his ears.

'Where am I, Vincent?' he asks.

'Hope I didn't wake you, Jim-boy?' Vincent says, turning to face him.

'Where am I?' Jim repeats.

'Wherever you want to be. There's a bloke down the corridor who thinks he's on a beach in Jamaica. "Pass the piña colada!"'

Vincent returns to work on the mirror. 'And the geezer next door thinks he's in Helmand, back with his old regiment. "IED! IED!"'

'What are you doing over there?' Jim manages to ask. It's not just his head. His whole body is in agony.

'Removing the mirror. Should never have been here in the first place. Not with a category one like you. But we're full to the brimmers, so needs must.'

'Why can't I have a mirror?' Jim asks.

'Because we don't want you giving yourself a fright,' he says, laughing. 'The sea wouldn't even give you a wave in your current state.'

Jim doesn't join in with the laughter. Instead, he closes his eyes and thinks again about his last conversation with Dr Haslam. *Paranoid schizophrenia...* Every volunteer reacts differently to incapacitating agents, which is why a large sample size is so important. It wasn't unusual for volunteers at Harwell to become psychotic.

'You're OK over there,' Vincent continues. 'Your leash doesn't reach this far anyway. But we can't take no chances with the mirror. Last month, a woman on the female wing nearly strangled herself to death with a bath plug,' Vincent says, pulling the chain out of the sink with a flourish. 'Ripped it right out and garroted herself. Month before, this bloke found some surgical gloves on a windowsill and swallowed them whole. They had to do the full tracheotomy treatment on him.'

Jim stares at the ceiling. Harwell used to be such a tightly run ship. Someone's losing their grip. Turn your back for a second and a volunteer's trying to fly out of a carelessly opened window.

'You shouldn't be here, you know, Jim-boy,' Vincent says. 'You're different. A gent. Things have gone from crap to shite since you left. People've started to go missing. Know what I mean? Something's going on. And the staff, they're...' Vincent wells up. 'I'm a cleaner, Jim-boy. Nothing wrong with that. Somebody's got to do it. But they're treating me like the shit I have to scrub off the walls. Vince do this, Vince do that. It's not right. Not right at all.'

'Dr Haslam called me a patient earlier,' Jim says. 'I'm not a patient, Vincent. I'm a successful government scientist, back at Harwell because I blew the whistle on what's going on around here.'

Vincent stops work on the mirror, now off the wall, and stacks it outside in the corridor. He comes back with a dustpan and brush and sweeps up the remaining glass shards from the floor beneath the sink. Once he's finished, he walks over to Jim.

'Who did you tell?' he says, sitting down at Jim's bedside like a doctor on his rounds.

'A journalist,' Jim says, suddenly overwhelmed. 'A beautiful journalist called Bella.'

Vincent nods. 'Think I might know her.'

'You do?' Jim asks. He hasn't got the energy to ask how on earth Vincent might know Bella. 'Went to Oxford, now working for a national newspaper. Tall as a giraffe.'

He remembers when she walked into the pub for the first time, how he'd seen her at the bar and had to look away, confused by her awkward beauty.

'That's the one,' Vincent says. 'Smart bird.'

'She's coming back here to Harwell to talk to Dr Haslam, get his version of events for her newspaper story. But she mustn't, Vincent. They'll do awful things to her. I've got to stop her.'

Vincent nods again and checks over his shoulder. 'So that's why I've left you a little present, in the sink,' he says, leaning forward to whisper in Jim's ear. 'I'll leave the door open too, as it's you, Jim-boy. Can't do any more than that. Haslam'll kill me otherwise. He's more dangerous than the patients. Proper psycho, that one. The short arses always are.'

But Vincent does do more. As he rises from the chair, he pulls out a key and unlocks the security tether that runs from Jim's tied hands to the bed frame, winking at him as he walks out of the room and down the corridor.

'Let them know, Jim-boy,' he calls out. 'You let them know.'

100

Bella

'You were keen on Wordsworth, very keen,' Dr Haslam says, looking at the spine of *Lyrical Ballads* as he closes it. 'Not many of you were.'

Bella hasn't come to the college library for a tutorial. She wants to know what happened to her friend.

'Shakespeare too. *And, like this insubstantial pageant faded, Leave not a rack behind. We are such stuff as dreams are made on—*'

'What did you do with her?' Bella interrupts. It feels wrong to be raising her voice in a library, but she's beyond caring.

'*...and our little life is rounded with a sleep.*' Dr Haslam pauses. 'Erin was very ill,' he says, glancing over to the door where she entered. Bella looks too but no one's there.

'Why didn't you let me see her?' Bella asks. 'I phoned, emailed, texted you so many times, but you never answered.'

'Shall we take a walk? In the gardens?' Dr Haslam says, replacing the book on the shelf. 'We light them up at night now.'

Bella's memories of this man were all from her last months here, when she felt herself, on top of her game. His smile was winsome, his beady eyes brimming with empathy and intellect. Now all she sees is a liar, a little man who's hiding the sordid truth from her.

'I'm not interested in going for a fucking walk,' Bella says, still thinking of Erin.

Dr Haslam glances up at her, a steely look at odds with the soft flesh of his face. She turns away. A wave of fear passes through her, shaking her confidence. She's back as the student and Dr Haslam is her tutor, a figure of authority who she's just sworn at.

'What did you say?' Dr Haslam asks, facing her now, pulling at the cuffs of his jacket like an agitated bouncer.

'I don't want to go for a walk,' she says, more tentatively this time.

'You'll do exactly as I ask,' he says, his own voice firm but measured. She remembers this tone, its undertow of menace, the effect it seems to have on her.

'I'm sorry,' she hears herself saying, staring at the carpet of the library. Why's she apologising? She looks up at Dr Haslam again, fighting back a tear. 'What did you do to her? To Erin? I just want to know what happened.'

Dr Haslam gestures for them to head for the door.

'As I say, she was very ill and, tragically, chose to take her own life,' he says, as they start to walk. 'As for what I "did" to her, we all did what we could to save her, the staff here and the medical team, but she passed away in hospital after a long fight, never regained consciousness. If I could have arranged for you to see her, I would, but she was far too ill for any visitors.'

So it's true, her friend is dead. She tries to keep it together as they reach the library door but she can't stop herself from crying. And wondering how Dr Haslam's version of events squares with the body that was found in the crop circle. Has she got it all wrong? Was it someone else? Another bird tattoo fanatic?

'Has there been a funeral?' she asks, stopping at the door.

'Not yet,' he says. 'She had no family.'

'She had friends,' Bella insists. 'People like me.'

'And you were a very good friend to her,' Dr Haslam says, putting a hand on Bella's shoulder.

Just not quite good enough. She recoils from Dr Haslam's touch, shaking off his hand.

'Erin was found in the middle of a crop circle in Wiltshire,' she says, holding on to what she believes to be true.

'As I understand it, they've not been able to identify any of those poor victims,' Dr Haslam says, his voice almost hypnotic in its evenness.

'The body in the second circle – the arms were covered in feather tattoos, just like Erin's,' Bella continues, fighting against the tone of control in Dr Haslam's voice.

'A lot of people have feather tattoos on their arms, Bella,' Dr Haslam says. 'They're quite the thing at the moment, so I'm told.'

'And she had a small rook tattoo hidden behind her ear – she showed it to me once.'

A rook that spoke to her when she danced. *One, two, three! One, two, three!*

'I wouldn't know about that,' Dr Haslam says. He seems genuinely surprised.

'You don't understand,' Bella says, insistent now. 'I'm

writing a story for a national newspaper about the deaths. I've spoken to a government scientist at Porton Down who knows what the circles mean. They're the formulas for chemical weapons that are being tested on innocent victims like Erin. And it's all happening at Harwell, down the road from here.' She thinks of Jim's diary, the multiple references to Dr Haslam. 'A place that I think you know only too well.'

'Harwell?' Dr Haslam asks, more amused than surprised. 'Can't say I've ever been there. But I've heard a lot about the place from others. The famous synchrotron. But listen, your visit here's most timely.' He turns to open the library door. 'We're running a residential course at the college, teaching English as a foreign language, and could do with some help. Your room's ready—'

'The fuck are you talking about?' Bella says, staring at Dr Haslam in disbelief. Is that why the light was on in her old room? 'I've left this place. I've got a life, a job as a journalist.' She relishes saying these last words. Dr Haslam never took her Fleet Street ambitions seriously, her desire to follow in her dad's footsteps.

'And where's Mum?' she adds. If there is something between Dr Haslam and her mum, he must know where she is.

'Your mum?' he asks, holding the door open for Bella as he follows her out into the corridor. 'Your mum's right here.'

Bella looks up to see her mother being ushered out of a side room by a porter, eyes blood-red with crying.

'Hello, flower,' she says softly, holding one of Bella's suitcases. 'I'm so sorry. I've brought your things.'

101

Silas

'I see they've brought the big red key,' Silas says to Strover, as they watch the emergency response team unload a bright red battering ram from a police van in the car park at Cranham Hall. 'Of course they have.'

Silas doesn't expect to use the battering ram but nor does he have any more keys if they discover another secret morgue.

'Remember, this is a psychiatric hospital,' Silas calls across to the officers, as they continue to get ready. 'So we go in nice and gently. It's not another drugs raid in east Swindon.'

He's done enough of them in recent weeks, targeting gangs in Walcot, Eldene and Park North. But unlike those dawn raids, there's no element of surprise here. Silas has already informed the manager that he's about to search the premises, mindful of the patients' well-being. The circumstances are not ideal – unannounced visits are always better – but at least it's dark. As a precaution, the uniforms on all exits will check everyone who enters and leaves the women's wing in the old hospital building at the front and the men's wing in the modern extension to the rear.

'Here come forensics,' Strover says, turning as a lorry enters the car park, its headlights sweeping across the lawn in front of the main house.

'About bloody time,' Silas says, glancing at his watch. They've taken longer than the promised twenty minutes to drive over from Oxford. 'Once we're in, I want you to focus on the women's wing and I'll look for Jim in the men's.'

Silas nods over to a BMW X5 in the far corner of the car park, where three officers from the Tactical Firearms Unit have assembled. 'And we only call in the cavalry if we really need it.'

'Am I looking for anyone in particular?' Strover asks.

Silas hasn't yet shared his theory with her, who else they might find inside the secure hospital.

'Take CSI to the mortuary and then start asking around, see if any patients or staff knew of a tattooed patient called Erin,' he says, pausing. 'And keep your eyes open for Bella.'

'Bella?' Strover asks, surprised, as her phone starts to ring.

Silas nods. Strover holds eye contact with her boss while she takes the call. She's always believed in Bella's involvement in the case and now she's about to be proved correct. He's not sure what he'd do without her. It didn't take long for the magistrates' court to fast-track a search warrant for Cranham Hall, thanks to her help with the paperwork. Getting a warrant for AP Brigham's headquarters in Reading required more persuasion, but colleagues from Thames Valley have already started to search the premises, looking for evidence of the three untested antipsychotics.

'That was the hospital,' Strover says, coming off the phone. 'Noah's regained consciousness.'

'He took his time,' Silas says, but he's pleased Noah's out of danger. He had no idea what he was getting into when Jed Lando commissioned him to make three crop circles. A gentle soul caught up in something horrific. Their visit to his remote farmhouse seems a long time ago now.

Silas looks over at the dark Victorian building and sees a solitary pale face lit up in a high window. If Lando was drawing attention to antipsychotics made by AP Brigham, how many Erins are there? How many victims died in their development? Two were found in the crop circles, but were there others? He wouldn't put it past the American CEO, last seen dancing with an inflatable syringe. And did they all die in illegal trials or did some survive, living a life of dopamine-fuelled delusion?

'OK, let's do this,' Silas says.

He takes a deep breath and sets off for the entrance gate with Strover by his side and Jim and Bella on his mind.

102

Bella

'You've got to tell her,' her mum says, turning from Bella to Dr Haslam. 'It's just not right.'

'What's not right, Mum?' Bella asks, increasingly desperate.

The three of them are seated at the end of a large oval table in one of the college's more formal rooms near the library. Dark, oak-panelled walls, a high white ceiling with ornate cornicing, an old crest of arms above a brick fireplace. This was only ever for the staff. Bella had assumed it was the fellows' dining room, or for 'viva' exams, which makes her feel even more uncomfortable being here now. Her mum is in tears but Dr Haslam is seemingly unmoved, eyes shifting from Bella to her mum and back again like a poker player. A porter stands guard outside the half-open door.

'If you don't tell her, I will,' her mum says, reaching across the polished table to clasp Bella's hands in hers.

Dr Haslam raises his own hand, as if he's stopping traffic. Bella watches as her fingers start to slide away from her mum's and she puts them in her lap. And then her back straightens and she's sitting upright like an attentive student.

'That's better,' Dr Haslam says quietly. He lets her mum's snivelling subside until there is silence in the room. His tutorials were often like this, she remembers now. A terrible quiet, broken only by distant, tormented wails.

'We all delude ourselves in life,' he says, getting up from the table to walk around. 'After a particularly good violin practice, I tell myself that I'm a worthy match for Yehudi Menuhin. And who is anyone to say otherwise? It's what keeps us going. The lies we tell ourselves.'

He stops behind Bella, hands coming to rest on her shoulders. She doesn't flinch this time, not even when he leans down to talk into her ear.

'You believe you studied here for three years, improved yourself, came to understand the works of Wordsworth and Shakespeare, had your awakening in the shadow of Oxford's dreaming spires. That's your reality. And it served you well when you went out into the world.'

He removes his hands and sets off around the table again, holding court.

'You know my thoughts on the media as a career, but you told yourself you were a journalist, just like your father, and look at you now. Here on your first story for a national newspaper. How wonderful!'

He waves his arms about theatrically, as if he's introducing Bella to an audience. And then he returns to his seat, milking the tension, before turning his eyes on Bella.

'But from where I'm sitting,' he says, 'from where your mother's sitting, none of it is true.'

The room falls quiet again, apart from her mum's muffled sobs.

'How do you mean?' Bella asks, turning to her mum for an explanation.

'He means you didn't...'

Dr Haslam interrupts her with a raised hand again, his fingers pale and clammy.

'But that's not to say your reality is devalued, just because it doesn't equate to mine, or to your mother's,' he continues. 'It only becomes delusional when you compare yourself to others. For you, it's the truth and that's all that matters. No one can take it away from you. And nor should they.' He throws a glance at her mum. 'It can be dangerous, wilfully trashing other people's private worlds, puncturing the bubbles they choose to inhabit, destroying what they believe in.'

'I studied here for three years,' Bella says slowly, her voice shaking. 'Three long, hard years,' she adds, stabbing at the table with her finger. She looks from Dr Haslam to her mum and back again at Dr Haslam. 'Are you trying to tell me I didn't?'

'Far from it,' Dr Haslam says, hands held up in innocent protest. 'It was an education – for all of us. You arrived a mess and left us as a together young woman.'

Dr Haslam has no idea how patronising he can sound.

'And I'm now working on a national newspaper,' Bella continues. 'As an investigative journalist.'

The room is starting to spin.

'Absolutely,' Dr Haslam says. 'You say as a journalist, your mother says as a secretary. Take your pick. I know which one I'd choose in your position. And that's what you've done. "Fake it until you make it" – isn't that what the young say these days? Good for you.'

Bella flashes a look at her mum, who turns away, embarrassed, in pieces.

'Is that really what you think I do at work?' Bella asks her.

A sudden commotion outside. Dr Haslam looks up at the door, concern on his face. The porter seems shocked too and disappears down the corridor.

Bella senses time is running out. She needs to remember why she's here. The room spins faster, forcing her to grip the sides of the table, close her eyes. She's come back to her old college for Erin.

'Erin's body was found in a crop circle in Wiltshire,' she says, fixing Dr Haslam in the eye. He doesn't like that, fighting fire with fire. This time it's him who seems agitated, troubled. More noise outside.

'I know it was her,' Bella continues, 'and I need to know why she got from here to there, how she died, what it all means. The patterns, the codes. And you're the one person who can tell me.'

As if on cue, a parliament of rooks rises up from the sycamores outside. Something has disturbed them. They fly around the treetops, urging Bella on with their hideous cries.

'She was my best friend,' she says, dizzy now from the spinning. 'We laughed and cried, here in this building where we studied together as undergraduates. Out there in the college gardens. Ornithology for her, English literature for me. She did too many drugs, everyone knows that. But she came from a broken home and found a family here. Friends.

And then one day she disappeared and I let her down. I didn't do enough to find out what happened, how she ended up dead on a hillside in Wiltshire. We were college friends but life moves on out there. People go their separate ways. It's no excuse, though. I failed her. But I'm back here now and I'm not going anywhere until you tell me what happened to Erin. Lil fucking crazy Erin.'

The door is flung wide open and Bella screams. Jim is standing there, his head bandaged and a deranged look in his blinking eyes. But he's not on his own. One arm is wrapped around a terrified woman's neck and in his other hand he's holding a shard of mirror, pressing its jagged edge against the soft skin of the woman's throat.

Bella screams again.

'Nobody move and everyone will be fine,' Jim says. Bella stares at him. No glasses. He can't see properly. What's he doing here? At her old college? His cheeks are criss-crossed with deep cuts and a patch of crimson is seeping through his head bandage. Bella glances at the woman, who's wearing a uniform of some sort. Who is she? The college nurse? Bella's seen her before somewhere.

'Jim,' Bella says, in case he hasn't spotted her. His eyes dart around the room and his nostrils flare. All his senses are heightened, assessing risks at a primal level. For a second, Bella thinks of Rocky, his chameleon, and imagines Jim's forked tongue lashing out at Dr Haslam.

'Put the glass down,' Dr Haslam says, in a tone that Bella has come to fear. Prospero talking to his Caliban. She knows that Jim won't be able to resist his words. No one ever can. But Jim just smiles, a bruised, painful smile. What a bloody legend. She just wishes he'd let the poor woman go.

'Have you given your side of the story, then?' Jim asks, addressing Dr Haslam, whose spell appears to have finally been broken. 'Told Bella about all the experiments that go on here?'

'What experiments?' Bella says.

Jim looks thrown by the question. Dr Haslam sits back, happy to listen.

'The ones I told you about,' Jim says. 'In the USB files.'

'But… they were all at Harwell,' Bella says. She looks around the room, rocking like a ship in a storm now, the floor heaving up and down. It smells like college, that weird mix of furniture polish and sanitiser, and the windows have familiar bars on the outside, just like those in her old room. What's Jim talking about?

'And?' Jim asks.

'We're at my old college,' she says, eyes widening in fear and confusion.

Jim shakes his head, his own eyes defiant with conviction. 'We're at Harwell, Bella. This is Harwell.'

'And I can play the violin like Yehudi Menuhin,' Dr Haslam says, standing up to face Jim. Her mum puts her head in her hands as Bella looks around for somewhere to be sick.

'Now, if you'd kindly let my duty manager go,' Dr Haslam continues, 'we can all sit down, get our quotes, talk about the experiments and behave like civilised human beings.' He pauses, fixing Jim in the eye. 'Let her go, Jim.'

This time, Jim is no longer immune to Dr Haslam's sorcery and begins to lower the sliver of glass, inch by inch, away from the woman's neck, his face full of disappointment at what his hand is doing. Dr Haslam reaches out his own hand but as his fingers circle the glass, a cry rings out from down the corridor, followed by the sound of running.

'Armed police! Armed police!'

Two officers in black helmets and goggles burst through the doorway and point their guns at Dr Haslam.

'Drop it,' the officer shouts, glancing back down the corridor. 'Drop it now.'

Dr Haslam lets the glass fall to the floor as is if he's releasing a potion.

'Everyone put their hands in the air,' the other officer says, his voice bursting with adrenaline. A third armed officer runs into the room, radio crackling.

'We have two casualties on the second floor, old building,' he says into his lapel mic. 'Request urgent medical support.'

'Oh Jim,' Dr Haslam says, shaking his head in disappointment. 'What have you done?'

'Shut the fuck up,' one of the officers shouts, pointing his gun at Dr Haslam's chest. Bella stares in disbelief. Her tutor – head of pastoral care at the college, a 'big don', as Erin called him, feared by all – has no sway over anyone any more. No one at all. All respect gone. The officers bark more orders into their radio mics as they check everyone for weapons.

Bella, arms above her head, feeling faint, is back on Studland beach.

Dad's dead, Bel, and he's never coming back.

The police had intervened then too, shouting at her as if she were a wild animal. Her mum had also screamed.

Let go of her, Bel! Bel! Please! Helen can't breathe!

Everyone was so angry with her. Did Helen die that day? Did Bella tighten her grip until her sister's eyes popped with fear and her breathing ceased? Christ, is that what really happened? Did she kill Helen? Is that why her heart's racing now? Or did she let her go and allow her to live? She flexes her sweating hands. If only she could remember.

Bella looks up as a man in a suit enters the room and walks over to Dr Haslam. Bella's vision is blurred but she recognises him as DI Hart, the detective who came to Jim's home. A familiar woman walks in behind him. DC Strover. She glances at Bella, a look of reassurance in her eyes.

'Dr Haslam, I'm arresting you in connection with the death of two psychiatric inpatients here at Cranham Hall,' the detective says.

His words are faint, drowned out by another police voice in Bella's head.

I am detaining you under section 136 of the Mental Health Act 1983.

'And for ordering the murder of your former colleague Jed Lando,' the detective continues, but Bella can only hear her mum now.

We just want you to get well again. They'll take good care of you, I promise, my flower. And once you're better, you can go to uni, get a job in journalism, make Dad proud.

'I'm also arresting you in connection with the death of five other inpatients as well as on suspicion of administering unlicensed drugs in contravention of the Medicines Act 1968.'

The detective glances at Jim and Bella, but all she can hear is Dr Haslam on her first day at college.

Welcome to Cranham Hall, Bella.

'You do not have to say anything. But, it may harm your defence if you do not mention when questioned something which you later rely on in court. Anything you do say may be given in evidence.'

Dr Haslam stares back at the detective, powerless, emasculated, a shadow of the man he once was. And now Bella understands, remembers his more recent words.

The college nurse is going to try you on some new medication, long lasting, a single depot injection. I think you'll do well on it, Bella. It should help you to concentrate better on your studies. And farewell to pills and weight gain and all those other nasty side effects. You might forget a few things about your time here with us but we'll have you ready for the real world before you know it. We'll miss you, Bella. Your love of reading, of Wordsworth and Shakespeare. Your search for the truth.

Outside, high up in the sycamores, a rook calls out in approval as Bella collapses unconscious into her mum's arms.

103

Silas

One week later

Silas squeezes Conor's shoulder and gives him a smile, watching as Mel leans over to kiss their son goodbye. He's doing well, responding to a mix of lighter medication and something called Open Dialogue, an alternative, more family-based approach to mental health issues developed in Finland. Jonathan, the psychiatrist, is a big fan.

He is waiting in the corridor and ushers them into a small side room, where Silas's eyes alight on a tray of dark chocolate digestives. It would be rude not to.

'Thanks for your time,' Silas says, tucking into a biscuit as they sit down. He expects Mel to remove the tray, but she leans forward and takes a biscuit herself, smiling at him.

He's in her good books again, not just because he turned up on time for today's appointment, but because he's responsible for Jonathan being back on Conor's ward. Once the American CEO of AP Brigham had been arrested, the company's aggressive takeover of the psychiatric unit in Swindon fell through, much to everyone's relief, including

Jonathan, who had been transferred to the north-east at short notice.

'He's doing really well,' Jonathan says. 'You should both be very proud.'

'Thank you,' Mel says. 'For all you've done for him.'

'Thank your husband here,' he says. 'For getting my job back – and for saving your son from a course of medication that sounded horrific, to be honest. There's no other word for it.'

A lot has come out in the media about the drugs that AP Brigham was illegally testing on unsuspecting psychiatric patients at Cranham Hall, not least because of the shocking death rate. Five more bodies were found, thanks to the big red key. Sure enough, they had been taken from Jed Lando's private morgue to another locked room in the basement.

As the associate professor of chemistry at Imperial had suggested, all three of the experimental drugs, which were administered by one depot injection every six months, blocked the brain's D_2 dopamine receptors, but at lower levels than existing second generation antipsychotics, thereby avoiding extrapyramidal side effects and other negative symptoms such as weight gain. To achieve this, however, not all the so-called positive symptoms – the hallucinations and delusions – were reduced.

Jonathan has already taken time to talk through the case with Silas, explaining about the medication. He's also met with Jim and Bella, offering to help in their recovery.

'Some delusions remained and were even heightened by these wretched meds,' he says, taking a biscuit too. 'They also seemed to increase the memory loss that's often associated with antipsychotics.'

The experimental medicines themselves – designed to get people living with schizophrenia out of secure units and into the community, at considerable financial cost, of course – were found in the raid on the Reading HQ and at a nearby laboratory, where the American CEO had squealed like a pig when he was arrested. Most importantly, he revealed the central role of his colleague Dr Haslam, the psychiatrist who oversaw Cranham Hall, and the information was relayed to Silas as he and Strover had entered the building in search of Jim and Bella. The CEO also admitted responsibility for the deaths of Dr Armitage and Jed Lando. The killer he hired – the fake doctor – was subsequently arrested in Cornwall.

As for Cranham Hall, an inquiry has been launched and the hospital closed down, its name joining a growing list of mental health institutions in Britain that have been exposed and shamed for abusing and mistreating their patients. Not all of the dead have been identified yet, but they appear to have been homeless: troubled souls who weren't missed alive and whose deaths went unmourned, let alone investigated.

Their passing would never have come to light if it wasn't for Lando, who had become increasingly concerned by his company's secret drug trials, so much so that he decided to blow the whistle on them by commissioning a series of cryptic crop circles. The corpses that he left in them and their grim condition – lobotomised, straitjacketed – more than achieved his aim of drawing attention to the grim side effects of antipsychotics. Whether it was a twisted act of redemption for past sins, no one will ever know. He might also have felt that he'd tried in vain to blow the whistle at Porton Down, thirty years ago, and this time needed to adopt a different, more dramatic approach. And then there was the last circle, containing his own zombified, cancer-ridden body. Silas will never forget that night when he found him in the moonlight.

'Jed Lando had a difficult situation on his hands,' Jonathan continues. 'Two of the more successful guinea pigs, Bella and Jim, remained in the grip of particularly strong delusions. Bella believed Cranham Hall was an Oxford college, which she attended for three years before becoming an investigative journalist – what we call a "grandiose delusion". Jim believed he was being followed by MI5 – a classic "persecutory delusion" – as well as working as a top government scientist at Porton Down, when in fact he was selling rabbits at a nearby pet shop.

'So what was Lando to do? It can be very dangerous to march in and disabuse someone of a psychotic delusion, insisting that they are mistaken. On the other hand, one mustn't foster the delusion either, as Dr Haslam appears to have done. It's a fine balance – personally, I try to respect a patient's very real emotional responses to a particular delusion rather than dwell on its content.'

'What did Lando decide on?' Silas asks, glancing at Mel. For once, she seems happy for him to be thinking about work when they're visiting Conor.

'I believe he gave them a gift,' Jonathan says.

'A gift?' Silas asks, surprised. He wouldn't like to be at Lando's house for Christmas.

'In both cases, their delusions were rooted in reality,' Jonathan says, indulging Silas's confusion with a smile. 'Bella had been offered a place to read English at Oxford before her breakdown. Jim had worked at Porton Down during two summer university holidays as an intern. Unfortunately, he had a psychotic episode sometime after he graduated and before he was due to start at Porton. His troubled brain processed his recovery at Cranham Hall as a three-year secondment to an affiliated site at Harwell, where he believed he oversaw – and was subjected to – chemical weapons experiments.'

'What was the gift, then?' Silas asks, still baffled.

'The crop circles. They gave Bella a genuine news story and Jim real evidence of unethical experiments – evidence that only he, the amateur mathematician and chemist, could decode. Never mind that he got the formulas wrong. Jim sees patterns and meaning everywhere – a form of what we psychiatrists call apophenia – and he was the only one to make the link between the circles and drug trials. Lando sent the pair on a journey that ended with them discovering their delusions for themselves.'

'He let them down gently,' Silas says, intrigued.

'Exactly,' Jonathan says. 'Jim was spot on about the bigger picture: people were still being subjected to horrific experiments in twenty-first-century Britain. Only they were taking place at an American-owned psychiatric hospital rather than a secret British military base. And Bella was right to investigate. Like her father, she had a nose for a good story. What she didn't know was that she was part of the experiment too. In effect, she was investigating herself.'

'Will they both be OK?' Mel asks, taking Silas's hand.

'With time and the right treatment, they should go on to live meaningful lives,' Jonathan says. 'Like Conor.'

Mel squeezes Silas's hand.

'And what if the experiment had been allowed to run its course?' Silas asks.

'A good question,' Jonathan says. 'They would have needed another depot injection after six months but I'm sure Dr Haslam would have found a way to weave that into the narrative of their fantasy lives. He was certainly keeping a close eye on them in the real world – he even had people posted outside Bella's newspaper office in London. Jim's delusions were also becoming increasingly persecutory and paranoid. Who knows what dangers they might have led him to? He already had a history of self-harming – trying to knock himself out. Bella's delusions were generally more benign but at some point she would have been found out. And the shock of someone else discovering that she'd spent three years at an acute mental health facility rather than an Oxford college might have triggered a far more severe psychotic episode.'

'Some might argue there's little to choose between the two establishments,' Silas says.

Mel nudges him in the ribs as Jonathan's face breaks into an uneven, mischievous smile.

'As a Cambridge man, I couldn't possibly comment,' he says. 'It was better that she found out herself, in the way she did – on a story,' he adds. 'There's also the fact that the meds seem to have triggered highly dangerous delusions in some of the other patients – the ones who died. It can happen – I won't go into the details, but take a look at dopamine supersensitivity psychosis when you can't get to sleep. Poor Erin, for example, had long thought she was a rook. But when she was given one of these experimental drugs, her delusions became more extreme and she actually thought she could fly. Tragically, she eluded the unit's shockingly lax safety measures and jumped off the main staircase at Cranham Hall.'

Silas closes his eyes. Erin was important all along. As was Bella. Strover had seen it from the start. He hadn't. He'll listen to his colleague more carefully next time.

'There's only one thing that I'm still unsure about,' Jonathan says, leaning back, hands behind his head as if he's on a sun lounger. 'Jed Lando trained as a pathologist, as we all now know, but he retrained as a psychologist in America, where he seems to have become obsessed with a rare condition known as *folie à deux*.'

Silas glances at Mel. He's told her about the words whispered in the night by Lando before he was killed in hospital and it's baffled him too.

'I think he couldn't resist the opportunity of putting Jim and Bella together, two intelligent misfits, to see if they bought into each other's delusions, which they did, of course,' Jonathan continues.

'The typed reader's letter he sent to Bella at the newspaper,' Silas says, turning to Mel. They'd matched the font with the old Remington found in his office at Cranham Hall.

'Exactly,' Jonathan says. 'Bella became more paranoid and Jim believed Bel was a journalist. But Lando also hoped that, once he'd brought the two of them together, their shared delusions would help them to discover the awful truth about AP Brigham's experiments – and about themselves.'

104

Bella

'You've got a visitor,' the nurse says, walking over to Bella. She looks up from the table, where she's playing Bananagrams with Jim, and manages a smile.

'Thank you,' she says quietly.

'You were beating me anyway,' Jim says. They are sitting in the family room of a progressive psychiatric unit in Oxford – not Cranham Hall, not her old college and certainly not Harwell Science and Innovation Campus. This place is bright and airy, all big windows and pot plants, and flooded with sunlight.

'And you spell "psychotic" with a "y" not an "i",' she says, glancing at his letters.

'I think I'll stick to sudoku,' Jim says. 'I've never been good with letters anyway.'

'Your diary was great,' she says.

He smiles awkwardly. 'Here's your visitor.'

Bella turns to see her mum and a man walking across the room towards them. For a second she doesn't recognise him but then, as her mum bends down to kiss her, she remembers.

'I hope it's not a problem,' the man says, hanging back. 'Your mother said you wouldn't mind if I didn't stay too long, tire you out.'

It's the news editor – at the newspaper where she was briefly a secretary, on work experience, a placement arranged entirely by her mum, Bella now knows.

'Of course, that's fine,' Bella says, blushing. 'Take a seat.'

In truth Bella's exhausted, more tired than she's felt in her life. She's been sleeping day and night since the shock of discovering the reality of the past three years. That Dr Haslam, 'English tutor and head of pastoral', was in fact a deeply fucked-up psychiatrist, and her Oxford college was a medium-secure mental hospital. That the incidents of Freshers' Week hazing she remembers, broken up by porters, was Erin being forcibly medicated by the hospital's control and restraint team. That her best friend's drug habit was anything but recreational, the heavy doses designed to dampen a once fiery spirit and a strong conviction that she was a rook. That her own 'final examination' was not a test of her knowledge of Wordsworth and Shakespeare, but a last medical check-up before she was released into the world, fuelled by an untested antipsychotic depot injection that nurtured delusions.

We are such stuff as dreams are made on.

Bella's exhaustion is partly her medication, but it's also the daunting thought of what lies ahead, the rebuilding of an entire life. A task not made any easier by her inability to remember much of her old one. Delusions and amnesia seem to go together, one feeding off the other.

'Why don't you and I go for a walk in the gardens,' her mum says, taking Jim by the arm and patting his hand.

Bella smiles as Jim is led away, leaving her with the news editor.

'I'm so sorry, I've just got to take this,' he says, his phone ringing. 'My boss.'

'That's OK,' Bella says.

She turns to look out onto the garden. Her mum's still feeling guilty about everything, keeps saying that she believed she was doing what was best for Bella. She'd also believed Dr Haslam when he'd reassured her that Erin was recovering in hospital. It was only when his lackeys bundled her into a car outside the house in Homerton and held her captive that she knew something was hideously amiss. Up until then, she had no idea that Bella was the subject of an illegal drugs trial – Dr Haslam had sugar-spun her daughter's recovery, her 'awakening', as an example of youthful resilience and determination, insisting that she mustn't do anything to puncture Bella's delusions.

She'd nearly blown it on the very first day, when she came to pick Bella up in the car.

'I so wanted to tell you,' she'd said on her last visit. 'You can't imagine how difficult it was for me. I had no one to discuss it with... I felt very alone. But I had no reason not to trust Dr Haslam. You were looking better than you'd been for years – in great shape. He'd turned you around. And I bought into the idea of one injection every six months with none of the awful negative symptoms. It sounded like a miracle cure, allowing you to live a near-normal life, and I so wanted it to work. For you, for me. For us as a family. But I guess in the end it was all too good to be true.'

It will take time but Bella is beginning to comprehend

everything that's happened. Each morning she wakes shocked all over again, but she's having counselling as well as visits from Jonathan, a psychiatrist friend of DI Hart's, who says her mum was put in an impossible position and was right not to rush in and destroy her daughter's elaborately constructed delusions.

Bella has forgiven her mum and she hopes others will too. She still believes she's got what it takes to be a journalist and her real Oxford college has been in touch, promising that a place remains open for her when she's ready – three years after she was first meant to go there. In some ways, there's not so much to get her head around. As Jonathan says, delusions are often no more than expressions of our deepest wishes – and of our most profound fears.

'Sorry about that,' the news editor says, coming off his phone to talk to Bella.

'That's OK,' Bella says. In truth, she'd forgotten all about him, lost in her thoughts.

'I wanted to thank you for that piece you sent us,' he says, sitting down opposite her and glancing at the Bananagrams game. There are a couple of sexual words in there – Jim's – and Bella finds herself blushing again.

'A bag of shite,' she says, remembering one of Erin's favourite expressions.

'Not at all,' he says. 'Clearly, some of the detail wasn't quite right – well, a lot of it – but the essence of the story was spot on. And very moving. Which is partly why I'm here. We've been looking into your friend Erin and managed to establish some more details about her.

'Sadly, it seems her mother's dead too and she had no other living relatives but we've pieced together her early life in Dublin and, with your permission, we'd like to run it – as a tribute to her – beside the main piece that will carry your exclusive byline.'

Bella's pleased. It turns out Erin's dad was a gifted musician, even had a record deal at one time, before his drinking ended his marriage, career and finally his life. When he died, that night in a tent in Dublin, Erin was sent to stay with an aunt in London, who developed a crack habit before also dying. Erin didn't have much luck when it came to family. After being taken into care, she ended up at Cranham Hall and never really stood a chance – cannon fodder for AP Brigham, who used 'expendable' people like Erin for more high-risk trials.

'I need to check with Jim, but all your quotes from him still stand, of course,' he continues. 'About the experiments, the hell he went through. We just need to change a few locations – Harwell to Cranham Hall – and the medications – chemical warfare agents to antipsychotics.'

'Nothing too major then,' Bella says, smiling again.

'A little bit of heavy lifting but the essence of the story is still true,' he says. 'The shocking reality of illegal drugs trials on unsuspecting victims.'

'I'd like to read the piece on Erin first, if that's OK,' Bella says. She'd neglected her friend when she was alive. The least she can do is look after her in death. Erin hadn't broken the spell of Bella's Oxford fantasy while they were at Cranham Hall. She was too preoccupied with being a rook. Everyone had their own issues and even if they were aware of other people's delusions, they seemed to accommodate them into their own – Bella thought Erin was studying ornithology, after all.

'Of course,' he says. 'And please feel free to add any anecdotes from college – from Cranham Hall.'

'It's catching,' Bella says.

'I'll email it all over for you to read,' he says, getting up to leave. 'Mark sends his best wishes, by the way. And he's looking forward to reading your "Overheard" column. He says he never actually commissioned it, but he'll happily run it – and pay any expenses, of course.'

Five minutes later, her mum and Jim come back into the room after their walk in the garden. They are getting on well in a way that makes Bella feel a little jealous, if she's honest. The small asides, exchanged glances. Maybe she just needs a break from all Bella's questions. The only topic they haven't tackled yet is Helen. Bella's tried, but her mum bursts into tears whenever her sister's name is mentioned.

'Are you sure she's not dead?' Bella asks, each time they meet.

She is struggling to know who or what to believe any more. So much of the past three years has proved to be untrue. Her unanswered letters and emails to Helen, the voicemail messages – what if they were simply the desperate actions of a deluded woman? Her nights are filled with dreams of Helen and they always end in the same way. Her sister lying dead at her feet in the soft Studland sand, Bella's red fingermarks imprinted on her broken neck.

It's time to ask again.

105

Bella

'How did it go?' Jim asks, sitting down opposite Bella. 'With that guy from the paper?'

'They're running my piece,' Bella says. 'As an exclusive.'

'That's great.'

Jim seems better after his walk. Her mum went off to talk to the duty psychiatrist when she came in from the garden with him, but she's been gone for a while now.

'They're also going to print a profile of Erin,' Bella adds. 'A tribute.'

Bella feels a sudden, overwhelming sadness for her friend.

'It's OK,' Jim says, putting a hand on hers. Jim played the piano at Erin's funeral, the Aria from Bach's 'Goldberg Variations'. His way of trying to make sense of her death. As Bella watched Jim play, lost in the music, humming to himself, she realised how much she loved him. The funeral ended with a live rendition of 'Farewell to Erin', a popular Irish reel that Erin used to step dance to on the streets of Dublin, leaving the congregation in pieces.

Bella looks up as her mum comes back into the room

to join them. Something's wrong. She's agitated, not her normal calm self.

'Is everything OK?' Bella asks, her stomach tightening.

'You've got another visitor,' her mum says, sitting down to pour herself a cup of tea. Her hand's shaking as she stirs in some sugar. She never takes sugar.

'The Pulitzer committee?' Bella says, trying to dissipate the tension. 'Here to present my prize for services to accurate journalism.'

'Someone a little closer to home,' she says, glancing at Jim, who looks over to the door, eyes widening.

Bella turns to see who it is. She blinks, twisting a fold of skin on her arm until it hurts, desperate to believe, to prove she's not dreaming. She's not. There she is in the doorway, more tanned than she remembers, carefree, the hint of bohemian that Bella could never quite manage.

'Helen,' Bella whispers, wiping away a tear as her sister approaches.

She's alive – and no one can tell her otherwise.

106

Jim

One month later

Jim watched Bella hug her sister that day and for the rest of the week, when Helen visited the unit every afternoon. He's never seen such overt displays of affection. Never had a sibling. Helen hugged Jim too, which was tricky – she's about half his height – but sweet of her. She's a good person, funny, more relaxed than Bella. She's also the spitting image of her mum and now back in Australia, having made her peace with her sister.

She'd left Britain after the attack in Studland three years ago, vowing never to return. And clearly fearful for her life. If it hadn't been for the Special Constable who'd prised Bella's hands from her throat, she would have died on the beach. During her three years at Cranham Hall, Bella had sent her hundreds of letters and Helen had read every one of them. Just as she had listened to all the voicemail messages and read Bella's emails. She had wanted to reply, to speak to her sister, but her mum had advised against it, as had Dr Haslam, until Bella was well again.

It's not been an easy month for Jim, despite the love

he's felt from Bella. He's not sure if he'll ever get over the discovery that he hadn't been seconded to a biosafety level four high-containment facility at Harwell but was, in fact, being detained in a high-dependency psychiatric unit in Oxford. That all the classified information he'd leaked about Porton Down was open source, widely available on the internet. That he'd been working in a pet shop for the last three months and not at The Lab.

It's obvious to everyone that Bella's recovery is going better than his. Even the arrival of Rocky, who has become something of a celebrity in the unit, has failed to help. With a bit of luck, though, today's outing will go some way to closing the gap between his fantasy past and future reality.

'You mustn't be too hard on yourself,' Jonathan says, breaking into his thoughts.

Jim's in an open-top sports car with the psychiatrist who has been working with him and Bella, and about to arrive at Harwell Business and Innovation Campus, sixteen miles from Oxford.

'It's going to take time,' Jonathan adds, as he drives up Fermi Avenue, the main artery of the campus. The car, a bright red 1970s Alfa Romeo Spider, is a little cramped for Jim and his knees are wedged beneath the glovebox.

'Where would you like to go?' Jonathan asks, as they both take in the futuristic surroundings. To their right, the shiny new buildings of the National Satellite Test Facility, where satellites will be put through their paces, and the Rosalind Franklin building, billed as the most electromagnetically stable place on Earth. To their left, beyond a grassy bank, the sleek profile of the Diamond Light Source building – the circular 'synchrotron', low and silver, like a grounded spaceship.

'Mind if we just drive around, see if the place rings any bells?' Jim asks.

'Of course,' Jonathan says, slowing to take a better look at the synchrotron. 'You sure you haven't been here before?'

Jim turns to him in surprise. 'Isn't that why we've come?' he says, managing a laugh. 'To find out?'

'I'm sorry,' Jonathan says, smiling. 'I mean, when you were younger, before you fell ill.'

Jim shakes his head. 'The first time I thought I had visited Harwell was on a three-year secondment.'

Jonathan has encouraged Jim to see his life in terms of before and after his first psychotic episode, to acknowledge the San Andreas-sized fault in his timeline that split his world apart.

'I did a bit of research,' Jonathan says, trying to keep the mood upbeat as they drive on. 'Harwell's all about space, health and tech these days. Any of those sound familiar?'

Jim shakes his head, sunk by a sudden wave of despair. They shouldn't be here. He's wasting everyone's time. But Jonathan doesn't say anything or pass judgement. He's good like that, lets things come to the surface in their own time. They drive on in silence.

'Mind if we head down there?' Jim asks, glancing at a tree-lined side road to their right. A building at the far end looks familiar.

'Sure,' Jonathan says, turning off Fermi Avenue.

Jim's wrong about the building. Jonathan drives on down the road, nursing the car around a corner, and pulls over. He lets the engine idle, sensing perhaps that Jim wants to talk. A delivery van is parked up ahead of them, across the road.

'I got a letter today,' Jim says, sighing as he looks up at

the blue sky, the contrails drawn across it like a grid for noughts and crosses. 'From The Lab.'

'The Lab?' Jonathan asks, killing the engine. Somewhere, the sound of birdsong. A blackbird.

'Porton Down. The Defence Science and Technology Laboratory. DSTL.'

Jim removes the letter from his jacket pocket, reading the 'care of' address of the psychiatric unit in Oxford where he and Bella are inpatients.

'DSTL used to have a small presence here, a while back, but not for a few years,' Jonathan says. 'I looked into it.'

'Because a part of you believed that I really might have worked here?' Jim asks, surprised by the psychiatrist's diligence.

Jonathan pauses before he answers. 'I just thought it might help you to understand what happened. The way the brain strives for meaning in the world when we find ourselves disorientated.'

'Could you open this for me?' Jim asks, handing over the envelope. It reminds him of the day he got his first letter from The Lab, offering him a summer internship. He'd let his dad open it at the breakfast table at their home in Swanage.

'If you'd like me to,' Jonathan says, taking it.

'I'm hoping it's an invitation,' Jim says. 'To resume my original job when I'm well enough.'

After Bella's Oxford college had honoured her original place, she'd encouraged Jim to write to The Lab, explain what had happened.

Jonathan opens the envelope and studies the contents.

'What do they say?' Jim asks, staring ahead. He doesn't

like the silence. Or the profile of another high-tech building up on the left that's caught his eye. Why hadn't he noticed it before?

Jonathan folds the letter away and hands it back.

'It's a rather formal reminder of your obligations under the Official Secrets Act, given the current media interest in your story.' He pauses. 'I'm sorry.'

The bluntness of the reply is not a complete surprise.

'I signed the Act when I was a student, on my first summer internship,' he says, thinking of the letter, his eyes still locked on the building ahead. What secrets do they not want him to share?

A moment later, a black Range Rover appears out of nowhere. Jim stares at it in disbelief before he turns away, shielding his face.

'You OK?' Jonathan asks, looking at Jim and then over his shoulder at the disappearing vehicle.

Jim nods, but he's far from OK. He just wants his dad to be better. And his mum to be alive too, if only that were possible.

'You're going to see a lot of black Range Rovers in the future,' Jonathan says. They've talked at length about what happened, how he was followed when he worked at the pet shop, not by MI5 but by people from AP Brigham, all of whom have now been arrested. 'Plenty of things that stir up bad memories. It's not easy, particularly with a brain as lively as yours, but not all the signs, the hidden messages around us, the codes in the sky and the sand and the sea – in the fossils – are for you. Sometimes you have to let them go, leave them for others.'

Jim agitates in his seat, uncomfortable, restless. Jonathan

has already mentioned a condition called apophenia, the tendency to perceive meaningful connections between unrelated things, patterns in random information.

'You want to take a walk around?' Jonathan asks.

'Sure,' Jim says, keen to stretch his legs.

'Watch out for security,' Jonathan says. 'There's a lot of it about.'

Jim smiles as he climbs out of the car, but not for long. A tall, bulky figure leaves the building up ahead and walks over to the delivery van across the road.

Vincent? The maintenance man? It can't be. He was arrested, like all the others. Jim hopes he'll be treated more leniently. The man opens the door of the van. Maybe he's on bail? He has the same prominent forehead, long face and big, looping gait. He's also whistling, like a blackbird.

Pass the piña colada!

Jim wants to call out, ask him about the high-containment facility, thank him for what he did that day, but instead he just watches as the man climbs into his van and drives away.

'Friend of yours?' Jonathan asks from the car.

Jim shakes his head and looks again at the building, squinting in the sunlight as he takes in its glinting edges and curves. It's set back from the road and security is tight, CCTV cameras covering all approaches. Two guards keep watch from inside the reinforced front door. Jim spots a bench, in the shadows to the left of the building, partially hidden from the road, and his piano fingers twitch. The grass is worn all around. A muddy place to sit in winter.

Get back inside.

'Won't be a moment,' he says to Jonathan, a sudden chill passing through him. He hops over a low wall and walks across to the bench, where he sits down, letting the surrounds sink in.

You'll have a broken neck if you don't get back inside.

Is this where it happened? Where he came for a break from his work that day? He rests his arms on his knees and lowers his head, trying to think, to remember, staring at the dried mud. A glint catches his eye. He bends down and removes something, prising it out of the ground.

'Can I help?' a voice says.

Jim looks up at a security guard, standing over him.

'Private property,' the guard continues, but there is no aggression in his voice.

'I'm sorry,' Jim says, standing up. 'My mistake.'

'Everything OK?' Jonathan asks, as Jim walks back over to the car, watched by the security guard, who is talking on his radio.

'All good,' Jim says, taking in the Harwell campus for one last time. 'We can go now.'

They don't say much on the drive back to Oxford. Jim asks to listen to Radio 3, which is playing *The Magic Flute*. He closes his eyes. Mozart's last opera is rich with numerological symbolism, full of threes. Three-part harmonies, the three women who serve the Queen of the Night, the three boys. And it begins and ends in E-flat major, a key with three flats. Stop it. He must stop it.

Jonathan reaches forward and turns the radio off.

When they reach the unit, Bella greets Jim in the family area with a kiss.

'How was it?' she asks, holding him close. 'Did it help?'

She looks across at Jonathan who has stood back, talking in quiet tones to one of the nurses.

'We need to visit your Oxford college,' Jim whispers, checking around him. 'Where you studied.'

Bella pulls away from him, searching his big eyes for an explanation.

'How do you mean?' she asks, glancing over at Jonathan again. He gives her a look of sympathetic concern.

But Jim can't find the strength to answer Bella's question, to go on living like this, trying to decode a universe he no longer understands. Instead, he looks around the family room, tired, despairing, and starts to sob in Bella's arms – hot, bitter tears for all that's happened in his life. For the mum and dad he has lost and the beautiful person he has found.

And for the excruciating pain in his right hand.

'You're bleeding,' Bella says, looking down at his clenched fist. 'What is it?' she asks, lifting up his hand. 'What happened?'

Jim uncurls his bloodied fingers as a nurse approaches. A small shard of glass is embedded in the soft palm of his hand.

'It's a piece of lens,' Jim says, lifting his hand up for the nurse to inspect. 'From my glasses.' He pauses, looking around at everyone. 'I found it at Harwell.'

After the sharp fragment has been removed and his wound dressed by the nurse, Jim and Bella talk all afternoon, interrupted by visits from Jonathan and the unit's own duty psychiatrist, who each spend time with Jim, letting him talk, assessing his mental health, his current levels of medication. There are no quick solutions, no easy ways to piece back his life, repair the shattered mirror. But talking to Bella helps more than anything. She understands. She's been there.

Knows what it's like when the train jumps tracks to run on parallel lines.

The shard of lens he found in the mud exists in two worlds. He gets that now. One in which it belongs to him and one in which it doesn't. It's up to him to decide. Just as it is with the Range Rover, with Vincent, the building at Harwell.

'Thank you,' he says, as the dying sun floods their favourite corner of the family room. Rocky's favourite too. 'Thank you for sitting at my table in the pub that night. For believing in me. For listening.'

But Bella doesn't answer. Instead, she takes Jim in her arms and holds him close until the sun finally sets and his burning mind has stilled.

Acknowledgements

I am indebted to my exceptional agent, Will Francis, and all at Janklow & Nesbit, in particular Kirsty Gordon. Many thanks too to the brilliant team at my UK publisher Head of Zeus: my peerless editor, Laura Palmer; her editorial assistant, Anna Nightingale; Lucy Ridout, whose structural edit went well beyond the call of duty; copy editor Jenni Davis; and Jon Appleton, who proofread the manuscript.

I know some people read the acknowledgements page before the book – a tempting crime that I've been guilty of myself – so the following will be necessarily oblique. Unfortunately I can't thank by name those who helped me with my research into Porton Down, the UK's secretive military research facility on Salisbury Plain, but I'm very grateful for the insights that I was given into 'The Lab' and its talented workforce.

The charity Mind and its excellent website (mind.org. uk) should be the first port of call for anyone interested in learning more about mental health, or for those seeking support in connection with any of the issues raised in this book. Its report, *Mental Health Crisis Care: Physical Restraint in Crisis*, was a truly eye-opening read.

The initial idea for the story of Bella and Jim came to me after I had the pleasure of interviewing Nathan Filer about his seminal work, *This Book Will Change Your Mind About Mental Health*, at the 2019 Marlborough Literary Festival. It's a powerful, lyrical piece of writing that introduced me to, among many other things, the campaigning work of psychiatrist Joanna Moncrieff and her own fascinating book, *The Bitterest Pills: The Troubling Story of Antipsychotic Drugs*. I also urge everyone to watch Eleanor Longdon's inspiring TED talk, 'The Voices in My Head'. And if you're on YouTube, take a look at Marcus du Sautoy's analysis of J.S. Bach's 'Goldberg Variations' in his lecture, 'The Sound of Symmetry and The Symmetry of Sound'. 'Novichok and Other Poisons', an article in the *London Review of Books* by Hugh Pennington, Emeritus Professor of Bacteriology at the University of Aberdeen, introduced me to puffer fish poisoning and tetrodotoxin.

In no particular order, I'd also like to thank Lou McGregor for her floristry tips; the Royal Literary Fund for their invaluable support; Dr Stephen Gooder for the chemistry lesson; Antonia Gooder for her African botanical suggestions; Tim Jones for his encyclopaedic car knowledge; Dr Andy Beale for answering my morbid medical questions; Mary Harper for sharing her formidable knowledge of Somalia; David and Janet Stock and her sister Christine for their Jamaican insights; Chris Stock for his army tales; Andrew Stock for the ornithological checks; J.P. Sheerin for the walking and Irish talking; NHS local hero Bruce Mason for his paramedic expertise; Gay Herd for BSO acoustic advice; Toby and Katie Ashworth for their loyal support; Tim Thurston for his interviews on Swindon 105.5FM; Gay Herd for and the Crop Circle Exhibition and Information Centre in Honeystreet, Wiltshire. Any mistakes are, of course, mine.

My interest in crop circles began in the summer of 2019, when my wife, Hilary, and I were on a walk with friends in the beautiful Vale of Pewsey in Wiltshire. Passing through the tiny hamlet of Honeystreet, we stumbled across a forty-strong Chinese film crew milling around the Crop Circle Centre, which was about to be opened by guest of honour Feng Shaofeng, a Shanghai-born movie star. It was a surreal occasion, to put it mildly, but Monique Klinkenbergh, the woman who runs the Centre, made everyone, including passers-by, feel very welcome. We found ourselves toasting the opening with champagne before Shaofeng disappeared in a helicopter to fly over a nearby crop circle for a Chinese reality TV show called *Life in Adventure*. Monique has done much to encourage better relations between croppies and farmers and the Centre is well worth a visit – for sceptics like me, who want to know how they are made, and for believers who already know.

Finally, I'd like to thank my ever-supportive family: Felix, Maya and Jago, who keep me young and help with the lingo (not a word they'd use); and Hilary, my first and best reader, without whose love, humour, endless patience and encouragement none of this would be possible. This book is dedicated to the memory of Stewart, her much loved father and my wise and wonderful friend.

About the Author

J.S. Monroe read English at Cambridge,
worked as a foreign correspondent
in Delhi, and was Weekend editor
of the *Daily Telegraph* in London
before becoming a full-time writer. His
psychological thriller *Find Me* became
an international bestseller in 2017 and,
under the name Jon Stock, he is also
the author of five spy thrillers.
He lives in Wiltshire.